THE DOWNING OF
TWA FLIGHT 800

THE DOWNING OF TWA FLIGHT 800

James Sanders

Zebra Books
Kensington Publishing Corp.

http://www.zebrabooks.com

ZEBRA BOOKS are published by

Kensington Publishing Corp.
850 Third Avenue
New York, NY 10022

First Printing: April, 1997
10 9 8 7 6 5 4 3 2 1

Printed in the United States of America

This book is dedicated to the 230 innocent people who died aboard TWA Flight 800.

CONTENTS

Contents

ACKNOWLEDGMENTS

To my wife, Liz, who was drafted into this fight. Females are supposed to be noncombatants, but the government decided to harass her because her husband was investigating them.

Bill Birnes, my agent. What began as an interesting project soon turned into a mission against abuse of power at the highest levels of government. An agent is supposed to get the best possible contract, then run for cover when the shooting starts. He stayed for the fight. Thanks.

Paul Dinas, editor-in-chief, Kensington Publishing Corp., had the same reaction to FBI abuse of the power with which it has been entrusted. CBS caved when threatened. Paul Dinas and Kensington Publishing stepped up to the challenge.

David E. Hendrix knew what he was getting into, but he did it anyway.

Mel Opotowsky, managing editor of *The Riverside Press-Enterprise*, and Norman Bell, assistant managing editor, who gave the approval to launch the story, knowing the storm it would create. Out west, journalists seem far less concerned with the wrath of the federal government.

Carolita Feiring, editorial art director, Bob Nash, copy editor, and Loren Fleckenstein, who spent countless hours shepherding the story through the maze.

My attorney, Jeff Schlanger, who stepped into the breach amid the exploding legal shells of an FBI salvo while I finished the manuscript.

Thanks to Liz's support system, Lee Taylor, Lucille Collins and TWA Norfolk agents.

Mark Sauter, a reporter with *Inside Edition*, who verified the NTSB document that plays a prominent role in the manuscript.

Billy Dale, a hero I've never met. The FBI and the Justice Department framed him on orders from the White House. His reputation was destroyed by unnamed government leaks reported in the press. He ran up $500,000 in legal bills fighting the might of an out-of-control government. Billy Dale was exonerated by a jury after visiting hell. He has been an inspiration.

To everyone in the NTSB, FBI, FAA, Justice Department, and Task Force who risked the wrath of the government: thank you.

Speculation that TWA Flight 800 might have been downed by friendly missile fire is outrageous, the FBI's chief investigator James Kallstrom had been saying since the first days after the accident. At first the missile story was only a bubbling of rumor, murmured in chat groups on the Internet and surfacing from time to time in letters in the news groups. But the story always appeared preposterous, even when former ABC News correspondent Pierre Salinger repeated it, and Kallstrom denounced the rumors that the 747 was brought down by an errant missile launched by a U.S. warship or aircraft.

"It's just not true," Kallstrom said. "It's an outrageous allegation." Investigators had gone to the highest levels of the Defense Department to rule out friendly fire as a possibility, Kallstrom said.

Pat Milton (AP), 9/16/96

CHAPTER ONE
ANATOMY OF A TRAGEDY

As the evening of July 17, 1996 began, Eastenders on Long Island's south fork had no idea that only a few miles away a joint naval task force was assembling for a critical test of a top secret weapons system. In towns like Westhampton, Mastic Beach, and along the Shinnecock Bay Inlet, as midweek parties began, as recreational boaters set out into the warm night, they could not have foreseen the light show that would soon light up the skies. At 2000 hours, July 17, 1996, a world away from the Town of Southampton's resort beaches, military zone W-105, thousands of square miles of ocean located south and southeast of Long Island, was activated by the United States Navy. Within minutes, from different locations around the sector, military activity increased as the various units participating in the operation deployed their aircraft and surface vessels. The 106th New York Air National Guard put a C-130 and HH-60G helicopter in the air. The Coast Guard cutter *Atak* patrolled just south of Long Island's Gabreski Air National Guard base, her sailors catching the last few rays of deep orange before the sun finally disappeared for the night.

Over the horizon, to the east, in zone W-105, U.S. Navy AEGIS guided missile warships prepared for the final evaluation of a multibillion-dollar upgrade to their software, radar, and Standard IIIA and IV anti-aircraft/antimissile missile. The AEGIS radar and target management system was the pride of the U.S. fleet, so powerful that Ticonderoga-class guided missile cruiser personnel were said to have bragged that a ship like the USS *Normandy* could single-handedly fight a nuclear war with a small country, and win. "AEGIS arrogance," they called it, a pride supported by the stubbing Tomahawk cruise missile tubes and the surgically accurate antiaircraft and antiship weaponry that bristled from the cruiser's deck. AEGIS warships protected the fleet and could fight battles on land, sea, or air, and in just a few short weeks, the USS *Normandy* herself would steam into the Adriatic to relieve the USS *Arleigh Burke* AEGIS destroyer and take up station to bombard the Bosnian Serb rebels with a barrage of Tomahawks.

But that was still months away. Tonight, the system itself had to be tested as the surface vessels sailed into position. At the same time, a Navy plane, with newly upgraded electronic equipment designed to work with AEGIS, slowly cruised. The plane was the key to the new top secret and highly complex radar tracking system that was in its third year of testing. The aircraft's onboard computer hardware, weighing 525 pounds, was the platform for a new software upgrade linked directly to the AEGIS warship's radar system. If all ran like clockwork, the computer link and integrated radar and communications net would make it possible for a defensive envelope to be extended more than thirty miles over the horizon even in the most dangerous of coastal battle theaters, despite the foulest of weather and the darkest of night. But would it work?

Zone W-105 was selected for this final precertification test because of the complexity of the area. It was as close to a simulated Persian Gulf environment as the Navy could get without leaving U.S. coastal waters. Long Island offered dense ground-clutter, and the constant flow of commercial air traffic out of JFK gave the navy the "neutral" radar blips it needed to test the discrimination skills of the targeting software. Meanwhile, navy planes were approaching the exercise area to present "friendly" electronic signatures for AEGIS to track and compute into the task force battle array. A "hostile" presence would soon appear in the form of a BQM-74E Navy drone missile launched in the vicinity of Shinnecock Bay, east of Riverhead, Long Island. The 106th New York Air National Guard and Coast Guard units would be "traffic cops" for the Navy drone as it briefly passed over land en route to zone W-105.

The drone, the friendlies, the neutrals, the task force surface naval vessels, the National Guard aircraft, and the interlocking radar were all part of a test of the Navy's new Cooperative Engagement Capability or CEC, and integrated radar network designed to be fully compatible with the Army's missile defense system in order to give the battlefield zone closest to the water comprehensive protection from cruise and ballistic missiles. The Army's antimissile development was controlled by a command called Force 21, with a headquarters at Fort Monmouth, New Jersey, just outside of Eatontown, near the Jersey shore. Attached to this Army program was a senior Navy officer named Admiral Edward K. Kristianson, whose expertise in computers and integrated data management system arrays made him the perfect candidate for senior-level liaison with the Army for this multibillion-dollar 21st-century warfare.

At about the same time as the naval units were heading into position, the gate agents for TWA's New York to Paris Flight 800 were announcing final boarding. As families said good-bye, fathers hugged their daughters, and husbands and wives promised to call one another as soon as the plane landed safely, the TWA cabin crew was checking seating assignments on the computer printout. Out on the tarmac, the baggage handlers were putting the last of the luggage aboard, while in the cabin, Captain Steven E. Snyder and his first officer Ralph G. Kevorkian completed their preflight checklist. Earlier that day, this Flight 800 had flown in from Athens and had to be cleaned, checked, put through maintenance, refueled, and resupplied for the return flight to Europe. The area around the huge 747-100 was like a small city as the ground crew fought against the clock to get the plane airborne on schedule. Even as children at the departure gate pressed their noses against the glass to watch the train of little baggage trucks wind away from the landing gear, no one could have known the fate that awaited Flight 800.

Not in their most terrifying nightmares could anyone, neither passengers nor crew, have conceived of the engine of destruction that was assembling itself just offshore, or of the resulting fireball that would consume everyone onboard when the plane's path brought it near the hot zone W-105.

For several days before the final test on July 17, an Army unit had been deployed at the Long Island site, participating in several training missions that included the launch of several drones. Shortly before 2030 hours on July 17, an all-clear signal was given to the drone's launch platform. No general aviation or commercial aviation traffic was in the area. It was safe. The missile launch unit fed in the trajectory instructions to the drone's computer and watched as the automatic launch sequence counted off to ignition. Within minutes of the all-clear, the drone was airborne.

At about the same time as the all-clear signal, Linda Kabot from Westhampton Beach on Long Island was snapping off party photographs at a Republican fund-raising event from an outdoor restaurant deck overlooking Shinnecock Bay. Linda was focusing her camera at the smiling faces of local Republican politicos and friends, not realizing that in the background high overhead in the purple sky, that little streak of light she'd seen would turn out in one of the photos to be an image of the Navy BQM-74E Navy drone, quickly descending to its altitude coordinates shortly after its launch.

In its preprogrammed trajectory, after the Navy drone reached its preset altitude, it then dropped to thirty feet above sea level and accelerated to more than 500 mph as it began a long left turn away from the clutter of Long Island's land mass. The drone settled into an east-southeast heading toward the Navy AEGIS surface task force cruising on station just over the horizon. As the missile shot through the darkness at the speed of an airliner, the passengers aboard Flight 800 were just settling into a routine in the minutes after their late takeoff. Seatbelts began unfastening as the cabin flight attendants began preparations for the long service through the night and into the breaking dawn over Europe, eight hours away.

High overhead at 20,000 feet a Navy P-3 Orion deployed from Brunswick Naval Air Station, Maine, turned its infra-red tracking system on as it assisted the hundreds of millions of dollars in Navy high-tech tracking equipment spread along the shore from Virginia to Long Island, installed to monitor the ongoing development of the Navy's CEC warship defense system. Tonight the P-3 would be part of the invisible eyes of the network, monitoring, along with the land-based equipment, every phase of what the navy expected to be a perfect shootdown of the drone missile already on its way into the heart of the AEGIS task force.

The Navy had invested a lot of money in the development of CEC, even before the disastrous Exocet missile attack on the USS *Stark* in the Persian Gulf during the Iran-Iraq war, when American warships escorted oil tankers up and down the Strait of Hormuz while under hostile Iranian shore batteries. Amid the flights of commercial airliners from both adversaries, U.S. and allied military aircraft, and hostile aircraft from Iran, it was next to impossible to discriminate between targets, neutrals, friend or foe. This heavy traffic, Saddam Hussein maintained, was how the *Stark* was attacked by his fighter pilots in the first place. It was part of the reason for the deployment of CEC.

Because of the complexity of modern electronic warfare, in which the front lines obliterate traffic of all types, CEC was designed to be an almost surgical radar tracking, evaluating, and targeting system which would make it possible for the Navy to enter hostile environments like the Arabian Sea. CEC could identify and track all commercial traffic, and friendly military surface and air traffic in and out of the countries bordering the Sea, while remaining on the lookout for a hostile cruise missile launch from any direction. The Navy believed this system would allow them to discriminate electronically among friend, foe, and background clutter and still fight a battle. At least that's what the Navy thought as their warships and planes glided into position on the night of July 17, turned their combined radars on, and began sweeping the area for the commencement of this final pre-certification test.

Deep inside the electronic brain of a second Navy P-3 working with CEC, the radar communications equipment in the plane linked to the AEGIS-CEC transmitted signals along a downlink to the vessels' AEGIS radar computers. They began to decipher images from among the land clutter, friendlies, neutrals, and the hostile BQM-74E Navy drone missile rapidly heading toward the task force. It was as if combined radars and computers suddenly took an electronic snapshot of the entire area and identified friend from foe while eliminating neutral aircraft. Then, almost instantly, the interlocking software of each AEGIS-CEC platform acquired the target drone, but were suddenly jammed by electronic interference. One radar broke through the interference, however, computed a shot through the thickening fog of multiple "hostile" electronic jammers, plotted its trajectory, and commanded the software to automatically select the platform best positioned to make the shot.

The computer software then launched a Navy Standard IIIA or IV antimissile missile, specifically altered to function with this new equipment, toward the oncoming drone. From over the horizon, no one except Navy personnel could see the whoosh of the rocket launch as the missile took off from its tube. The antimissile missile climbed high into the evening sky and rocketed west in the general direction of the low-flying cruise-missile drone, toward a position where its onboard computer was expecting to receive a midcourse correction. This signal was supposed to fine-tune the Standard missile's trajectory in order for the onboard semiactive radar homing device to lock onto the target as the Standard missile began its plunge toward the drone a few thousand feet below. At least that was the plan.

Commercial planes rising into the sky from JFK were unwitting participants in this final test of 21st-century technology. As TWA Flight 800 climbed towards 14,000 feet, heading eastbound over the water for Paris, it was about twelve miles off the south coast of Long Island over the horizon to the west of the military exercise as it crossed into the warning zone and technically became a "neutral." At the same time, the electronic receiver onboard the Navy Standard missile began sweeping its secure radio frequency, waiting for the course correction commands from the AEGIS computers to direct the weapon, now at its predesignated point, to where it was supposed to attack its prey.

But prior to the mandatory midcourse correction, the last AEGIS-CEC radar still tracking the missile and the drone through the heavy electronic jamming, suddenly went completely blind. The drone and Standard missile could not be tracked. In two earlier tests, all but one radar had been put out of action by electronic jamming. On July 17, the Standard missile was no longer under the control of the AEGIS-CEC system. Following its internal programming, it continued on its westerly course at 3000 feet per second, actively searching for a target.

In an instant, the Standard's internal radar acquired TWA Flight 800 at well above and to the west of the target drone. The antimissile missile's radar turned sharply to the right, aimed its inert warhead at the 747, and painted an electronic bull's-eye on an area just in front of the right wing. The missile leveled off in a direct line to its impact point, and then at full speed slammed into the fuselage several feet below the passenger cabin.

There was no instant explosion, as the dummy warhead missile sliced through the huge plane a sheet of paper, depositing a trail of reddish-orange residue in its wake. It roared through the fuselage and exited through the left side of the plane, just forward of the left wing, where it left a hole large enough to walk through. After the missile exited, passengers, seats, galleys, food carts, and suitcases were sucked out of the interior through the hole in the left side, leaving a 4700-foot trail of debris along the sea bottom during phase one of the three-phase breakup.

On its way through the interior of the 747, the missile seriously weakened the front of the nearly empty center fuel tank. The plane went into a dive, and eight seconds and 4700 feet after the initial missile impact, a small explosion occurred, beginning approximately in the middle of the center fuel tank. The top of the fuel tank bowed upward, but at this stage of the breakup, did not rupture. This caused the floor of the passenger cabin also to bow upward, breaking loose seats in the center rows 21, 22, and 23. The explosion completed the separation of the front of the plane from the fuselage, initiating phase two.

The force of the explosion followed the path of least resistance: forward, blowing out the weakened front of the center wing tank. The explosive force caused the forward fuselage to separate a few feet in front of the missile's path, where the fuselage had been greatly weakened. This blast propelled row 15, seats 1, 2, and 3, about six-tenths of a mile to the left, while a large piece of the fuselage above the R-2 door sailed six-tenths of a mile to the right. The front end tumbled end over end off to the left as the remaining section of the plane continued on in a steep dive.

The pilotless stump of the 747 began to roll to the left until the left wing tip pointed toward the water below. The fire from the center wing tank spread rapidly up the right side of the fuselage and right wing. At about 7500 feet the inner right wing tank exploded. The engines and about ninety-eight percent of the center wing tank came to rest on the ocean floor more than 12,000 feet east of the missile's point of impact.

As quickly as it happened, it was over. Flight 800 was gone, spread across the water in a flaming swath. Moments earlier, a Long Island FAA radar technician staring into his electronic view screen thought he had seen something approaching TWA Flight 800 just before it disappeared from the radar. He saw "conflicting radar tracks that indicated a missile." Then he filed his report and the paper trail had begun.

A short time after the incident, the White House Situation Room was advised that preliminary assessment of FAA radar data indicated that a missile had shot down TWA Flight 800, en route from JFK to Paris with 230 passengers aboard. By 2 A.M. on July 18, key federal intelligence and investigative personnel were informed via White House teleconference that TWA Flight 800 was brought down accidentally by a friendly missile during a Naval exercise. They had on their hands, they were told in the blamelessly antiseptic world of military corporatese, a "situation." The Department of Justice Command Center and FBI Strategic Information Operations Center also came to life as Flight 800 information began to trickle in. Each was connected by a video teleconference system (VTS) to the White House Situation Room.

The initial talk in the room focused not on a bomb, but on a missile. Some eyewitnesses thought they had seen something bright arching toward the jet just before it blew up. At the next video conference, about dawn, an FAA representative said there was indeed a "strange radar blip."

But there were far too many people crowding into these teleconferences to let the missile analysis stand. So word was put out that "at air traffic control on Long Island, FAA officials reviewing radar tapes were unable to find even the mysterious blip."

The radar tape did not remain on Long Island for long. It went to the FAA Technical Center in Washington, D.C. The FAA Technical Center team, headed by the FAA's Tom Lintner, concluded that there was an "unexplained blip" on the radar tape. U.S. military missile experts told the FBI that a missile with a semiactive radar homing system would show up on an FAA radar set in transponder mode, but that a shoulder-fired heat-seeking missile would not.

According to *Newsweek,* writing in the aftermath of the crash, when news of the disaster had been reported in almost every newspaper and magazine in America, the possibility of a missile bringing down Flight 800 was the topic of conversation at the 6 A.M. VTS tele-conference. They said that the Stinger theory—the Stinger is a shoulder-fired, American-manufactured missile—resonated with the FBI, which had picked up intelligence that some terrorists had been shopping for the lethal weapons. As the 6 A.M. meeting got underway in the VTS room, there was "a lot of breathless talk" about attacks by missiles, or MAN-PADS. Still, some experts were dubious. The Stingers handed out to the Afghan and Paki "muj" by the CIA were at least a decade old, and probably junk by now. The Pentagon cast further doubt on the Stinger theory with some simple math. The effective range of a Stinger is just over two miles, and its sensor can't lock on aircraft much above 11,000 feet.

The Pentagon had a growing problem. They had temporarily halted the CNN nonstop coverage of a possible missile by having a high-level source "leak" disinformation, that the blip was an "anomaly," which CNN then authoritatively passed on to the public. But now they had the potential for a bureau-cratic leaking sieve if they didn't get the missile talk under control. So a coordinated program of leaks began to appear, and gradually began to neutralize the few clues being unearthed by a few intrepid reporters.

Whether the president or vice president actually knew about these events in the hours immediately following the crash or even whether they knew about the cause of the crash itself is a matter of conjecture. Nevertheless, somewhere within the topmost eche-lons of the military establishment, whether it was for national security or purely political reasons, a cover-up was initiated to conceal the real cause of the crash from the American people. Maybe the identitites of the ships in the task force had to be hidden. Maybe it would be too embarrassing to reveal with the Democratic convention only a few weeks away. We only know that the true details of some of the most critical evidence assembled on the floor of the hangar at Calverton have never been revealed to the public.

This cover-up would have been easier to maintain had there been no witnesses. But witnesses were everywhere, and they had to be discredited or dismissed. Thirty-four civilians at various locations along the flight window across Long Island saw the missile rise out of the ocean and intercept Flight 800. After extensive FBI and military debriefings, these thirty-four people were found to be highly credible, too credible to be dismissed as flaky. Each, from a different location, had seen a missile exit zone W-105 and intercept Flight 800. For example, an on-duty Air National Guard pilot saw a missile going from east to west slam into TWA Flight 800. The Air National Guard put out a press release the next day saying only that an unknown object, going from east to west, was seen by the captain. A woman on a boat south of Long Island was taking photos while facing the east. One of the photos shows a missile exhaust trail rising out of zone W-105.

The missile itself had left tangible evidence of its flight path through the aircraft in the form of a solid fuel residue deposited on the seats in rows 17, 18, and 19. It also left a red trail attached to airplane parts that fell off into the ocean during the first eight seconds of the plane's breakup.

Knowing that the United States Navy had shot down the plane with a missile, a plan of action was developed to remove evidence from the scene that would implicate the United States Government. Coast Guard MPs from a closed facility with only a skeleton crew maintaining and guarding it were brought over to guard a dock when sensitive debris was brought to shore.

One source described a Coast Guard MP team that observed this happening. The story is partially confirmed by New York police officers who observed a highly sensitive diving operation in the Red Zone during the first days after the crash. They were prohibited from the area during the multi-day course of the operation. Debris was brought to the surface and placed on the boat, but it did not go to Calverton hangar, they said.

But the source of the Coast Guard MP team story went further. He said that not only was the recovery of airplane wreckage a clandestine operation, but that the team was debriefed by intelligence personnel —they identified themselves as CIA—and warned that anything they said to the media or to any other sources would be a violation of national security and that they would be punished accordingly.

A team of Navy divers was brought in to dive in a particularly sensitive area of the Red Zone. No divers from any other organization were allowed to approach this area.

The Navy divers brought debris up and placed it on a ship which delivered the cargo to the dock guarded by the Coast Guard MPs, who watched as missile parts were off-loaded and placed on a truck. Pieces of the 747 that had red residue attached were also loaded onto the truck, which then drove off to an unknown location. Unbeknownst to those charged with removing the evidence from the crime scene, they missed some of the reddish-orange residue.

On August 3, 1996, a seat was recovered from the ocean floor with a significant amount of reddish-orange residue attached to its back side. Over the next few weeks, as the seats in rows 17, 18, and 19 were recovered, FBI investigators at the Calverton hangar saw the residue trail extend entirely across the cabin, scorched into the backs of most of the seats in these rows. The FBI took five samples of the reddish-orange residue for analysis. But, once tested, the results became part of a criminal investigation and the FBI declined to release their findings.

As the cover-up moved forward, it took the form of a lengthy process of creating new truths while systematically hiding the evidence. A series of nightly leaks to the press by unnamed government "sources," the content of which became increasingly illogical, kept conditioning the American population into believing whatever the NTSB suggested. Ultimately, they settled on a "mechanical" finding. But the real cause all along was a terrible lack of judgment on the part of the Navy, who had used innocent civilians as human guinea pigs as they rushed a multibillion-dollar weapons system into its final certification test before it was ready.

CHAPTER TWO

THE VICTIMS AND THE AFTERMATH

Here is the NTSB preliminary version of what happened in the Flight 800 crash:

On July 17, 1996, about 8:45 P.M., TWA Flight 800, N93119, a Boeing 747-100, crashed into the Atlantic Ocean off the coast of Long Island shortly after take-off from Kennedy International Airport. The airplane was on a regularly scheduled flight to Paris, France. The initial reports are that witnesses saw an explosion and then debris descending into the ocean. There are no reports of the flight crew reporting a problem to air traffic control. The airplane was manufactured in November 1971. It had accumulated about 93,303 flight hours and 16,869 cycles. On board the airplane were 212 passengers and eighteen crew members.

The airplane was destroyed and there were no survivors.

But the real event was much, much more than the preliminary NTSB report suggested. Flight 800 was a pressurized tube of 230 people climbing through the atmosphere to a cruising altitude of about 35,000 feet, en route to a Paris vacation or a European layover. Liz Sanders's former trainee, Jill Ziemkiewicz, was working that flight.

Jill Ziemkiewicz was a new hire. With only two months seniority as a TWA flight attendant, Jill spent twenty days out of each month on reserve, waiting for the phone to ring. Each phone call brought a new experience, working as the junior member of a crew working a flight to somewhere she had never been.

When the phone rang earlier in the day on July 17, 1996, she was thrilled. This was her first assignment working a trip to Europe. Even better was Flight 800's destination: Paris. She called her mother an hour before leaving, then went to Hangar 12 at JFK Airport to leave notes in the mailboxes of her TWA Flight Attendant Academy classmates: she was going to Paris. She was working an international flight.

Jill caught a red TWA shuttle bus and rode with some of the crew over to Terminal 5. At the crew briefing she soon learned that she was the junior person. No surprise there. Dan Callas had been with TWA two months longer than Jill and he was immediately senior to her. After that, the seniority curve was steep. Third junior flight attendant was Ray Lang, with twenty years' experience.

It was a short walk from the briefing room to the departure lounge at Gate 27, where outside the wall-sized windows the giant 747 sat on the tarmac ready for the crew's arrival. A few passengers were already waiting at the gate as the cabin crew swiped their cards through the security lock, waited for the green light, and boarded the plane through the jetway. Once inside the cabin, they stowed their personal gear and prepared to receive the passengers who were already lining up for preboarding. Nothing out of the ordinary. Nothing unusual. Jill was filled with anticipation because she was going to Paris.

Jacques Charbonnier was the flight service manager. A French paratrooper who saw action during the Algerian civil war, Jacques was in his thirty-sixth year with TWA. He and his wife of twenty years, Connie, always worked the Paris flight together. In fact, they met on Flight 800 in the 1970s.

Marielsa Ruiz was assigned to Flight 800 with the Charbonniers. After the crash and loss of her friends, she remembered them and other members of the flight crew whom she had come to know over the years.

"In July," she said to friends, "I held Flight 800 along with Jacques and Connie Charbonnier, Janet Christopher, and many friends who would later make up the crew of 800 on July 17. Due to family reasons,

I was not on board that terrible night. On July 11, on what would be my last flight with the ill-fated crew, I worked first class with Connie and Janet. We had a passenger, a young woman, who early on had confided to Connie that she was afraid of flying. When, halfway through the service, we hit turbulence, the woman panicked and ran up to the galley visibly shaken. She said she was too scared to sit and could only feel safe by staying close to us. Even though we were full and very busy and the passenger was clearly in our way, Connie welcomed her to stay and proceeded to calm and reassure her with all the understanding, kindness, and compassion she was capable of."

A second crew boarded Flight 800. At the last minute their schedule had been changed. Originally assigned to fly to Italy and work a flight back, they were reassigned to cross the Atlantic on Flight 800 and work a flight back from Paris.

Almost 200 passengers boarded the plane. Sixteen were teenagers from Pennsylvania, Montoursville High School's French Club, who were escorted by five chaperones. Forty-seven passengers were under the age of twenty-one.

Michel Breistroff stopped at a pay phone in Terminal 5 and called his girlfriend of two years, Heidi Snow. He proposed and she accepted. He then joined the line of passengers boarding Flight 800 and found his way onto the plane.

Soon, like Michel, most of the passengers at the

departure gate had made their way along the jetway and gotten seated. But Flight 800 didn't leave at its scheduled departure time. Weather in Chicago delayed flights out of O'Hare and other airports that connected to Flight 800, so the plane remained at the gate waiting for the connecting flights to arrive. Some of the new passengers were getting nervous, anxious to be on their way, but the cabin crew settled them down and people relaxed. Even the cabin attendants stood down from the protocols of imminent departure and simply waited for the news from the gate managers that the late flights were finally coming in.

As time dragged on, flight service manager Jacques Charbonnier walked back out of the plane to a crew smoking area, sat down, lit up a cigarette, and re-hashed old times—flying the international routes in the '60s and the '70s—with flight attendant Ray Lang, who also took out a cigarette and lit up. Soon they were notified that the Paris-bound passengers from the Chicago flights had finally arrived and they had received the go-ahead to depart. They both got up, snuffed out their cigarettes, paused at the doorway, waved "so long," and went down the jetway together. A few minutes later, the jetway pulled back and Flight 800 taxied toward the runway.

Air traffic behind Flight 800 was getting heavy as Captain Snyder received his clearance from the tower to take off. As the passengers put their heads back and flight attendants strapped in for takeoff, Captain Snyder and First Officer Kevorkian throttled up, sending the 747 roaring down the runway, and pulled back on the wheel, lifting the plane effortlessly into the evening sky. Behind them on the ground, tower flight controllers picked the plane up on radar as it climbed towards its cruising altitude before leaving the eastern shore of Long Island. About fifteen minutes later, as the plane crossed an invisible electronic demarcation line in the sky, a radar screen miles away and below the event, recorded the entire sequence of the disaster on tape.

The "impossible" had happened. From decks of restaurants, boats floating out in the harbor, the cockpit of a National Guard aircraft, and even the flight cabin of another airline passing through the area, independent observers, most of whom were more than credible, saw a streak of light illuminate a patch of the nighttime sky shortly before they saw the huge fireball overhead. When news of the crash broke on the local news, many of these observers called authorities with the information that they had seen a streak of light approach the plane before it exploded.

Officially, Flight 800 had disappeared from radar just out of JFK and was reported as missing. Then, only minutes later, the organizational wheels began to turn, initiating the crisis appartus every airline has developed but does not ever expect to use.

Voice mail was issued to TWA officials, notifying

them of the crash. At 2105 hours East Coast time, Staff Vice President for Flying Rex Pitts, activated the Emergency Command Center in St. Louis. Twenty minutes later, TWA President Jeff Erickson was contacted in London. Ten minutes later, TWA Trauma Center personnel arrived at Lambert Field, St. Louis, to arrange escort assignments and travel to JFK.

At 2145 hours, TWA's trauma "Go Team" was activated and the professional, but always gut-wrenching process of managing disaster and its victims had begun. Soon, team members experienced in handling the worst of news—telling people that their loved ones had been killed—were dispatched to JFK.

At 2315 hours, TWA held its first press conference in JFK International's Building 14 press room, announcing that Flight 800 had gone down off the south shore of Long Island, that rescue efforts were already underway, but that there was no estimate on the number of possible survivors. No, TWA answered the press, they did not have a cause for the crash but NTSB investigators were also on the way and as rescue efforts continued, they would try to learn the cause of the accident. No, they didn't have any immediate information about the cause except to say that Flight 800 was down, that deep-sea rescue attempts were underway in an area about five to ten miles offshore, and that Suffolk County authorities were already on the scene, along with private boat owners and Coast Guard Air Sea Rescue vehicles.

At about the same time that Flight 800 news began

breaking locally, New York Mayor Rudy Giuliani arrived at Kennedy to console any family members who were there, consult with New York and New Jersey Port Authority personnel, and promise to commit resources of NYPD rescue units to the recovery operations on Long Island. Mayor Giuliani could only confirm that nobody knew much about what had happened, but that he was there to make sure the families were taken care of and to extend the responsibility of New York City, where the flight originated, to any areas where it could help.

At 0215 hours on July 18, TWA Vice President of Flight Operations, Rich Roberts, landed at JFK from Washington, D.C., aboard a chartered jet. There was very little that Rich Roberts knew for sure at that point except that rescue boats and helicopters were on the scene off Long Island and that private boaters had been maneuvering through a sea of flaming jet fuel to locate any survivors or bodies. But that was all he knew. It was pitch black in the crash area and, as is typical in situations like this, rescue operations were hindered by the darkness. Roberts was briefed on TWA's efforts to that point, but, whether due to a misunderstanding regarding TWA's Flight 800 passenger manifest or absolute chaos at the airport, apparently did not communicate the information that Mayor Giuliani had been expecting. Giuliani chastised the airline for being less than efficient about disclosing the passenger list to families of potential victims, and the result was an uncomfortable, if not downright ugly, political scene at the airport at the beginning of what would turn out to be a long and arduous ordeal for family members and TWA personnel alike.

At 0500 hours, just as first light was beginning to break over the horizon off the Montauk lighthouse on the eastern tip of Long Island, the TWA St. Louis–based Trauma Team began to set up operations at the JFK Ramada Hotel. A little over two hours later, at ten minutes to eight on the morning of July 18, the National Transportation Safety Board was officially advised by TWA that "Flight 800's passenger list has been verified." With NTSB approval, the passenger list was turned over to the Trauma Team to begin the lengthy process of next-of-kin notification.

TWA chairman Erickson arrived from London at 0945 hours. Ninety minutes later he held a press conference to publicly announce that the passenger list had been verified and confirmed and turned over to the NTSB. At the same time the next-of-kin notification was underway. The last family would be notified that evening at 7:25 P.M. Forty-four minutes after that passenger list would be released to the press.

Because passengers' families lived in Europe as well as the United States, the Trauma Team's operation would have to be international in scope. The team's 675 volunteers, employed by or retired from TWA, converged on the trauma headquarters set up at JFK and in Europe throughout the day on July 18 under the direction of Johanna O'Flaherty who, four years earlier, had begun to plan for how she and her team would operate in just such an eventuality. It was planning she hoped she would never need, but when news of the Flight 800 disaster reached her, she was glad that she had contingency plans in place.

O'Flaherty's plans called for each volunteer to receive an initial full day of training covering "how the operation is implemented logistically and how the teams are organized." In her plan she wrote that "Annual recurrent training is also required of all team members. They have also had training from the Army's Department of Casualty and Memorial Affairs, which has briefed them on dealing with all the very sensitive issues surrounding fatalities; psychologists have spoken to the teams on stages of the grieving process. Additionally, the team has been trained as to how to deal with the media politely and firmly, attorneys and insurance issues, medical examiner's office, and obtaining dental records."

Most people never come into contact with corporate or government trauma teams, and that, of course, is a good thing. Therefore, at least in disaster situations, most people don't realize the extent of training and hands-on practical experience the trauma team's members usually bring to victims and survivors. The trauma team's work "runs the gamut from the almost mundane and routine immediate needs of transportation, child care, food, and clothing for the families and extended family members" to "working with the medical examiner's reports to assist in confirming the identities of the dead victims." It's a difficult job, especially when you're a member of a team from the airline whose plane has gone down, resulting in the tragedy in the first place.

At the same time, members of the TWA Trauma Team were victims themselves, because of the number of TWA personnel who were killed when Flight 800 went down. This was largely overlooked by the media—indeed, it is often an overlooked fact when an air disaster happens, that the members of an airline flight crew are usually well-known to trauma team members as well as to other airline personnel who are called into service during a disaster. On aircraft like the 747, which usually carry a full complement of flight personnel, a significant percentage of victims will be airline employees. Therefore, what has been largely invisible to the press and general public is that fifty-four members of the TWA "family" were lost on Flight 800, almost twenty-five percent of the total casualties.

In the next few days, as the trauma team began its work with the victims' families and the media converged on the disaster team headquarters at JFK, as well as on the recovery operations sites on Long Island, the full magnitude of the crash began to dawn on the world. Whatever the cause of the plane's going down, it was apparently a massively catastrophic event that caused a breakup while the plane was still in the air. While some people speculated about a terrorist bomb, perhaps planted to coincide with the Olympics in Atlanta, and others about a systems failure in the aging 747 that had gone undetected by maintenance crews, the real truth of what had happened was itself systematically hidden from the press under the camouflage of rescue operations.

While the cameras were focused on the floating wreckage and huge pieces of aircraft brought up from the bottom on cranes, the critical evidence, metal parts from the Navy Standard missile and exterior pieces of residue-streaked airline fuselage, had been removed in the predawn darkness to a location other than the Calverton hangar. Thus, while NTSB personnel began the next to impossible job of assembling a jigsaw puzzle and working backwards to ascertain the cause of the downing of Flight 800, the critical information had already been removed. No one would know about it, and the disparate pieces of eyewitness information, video records, and even swatches of seating material with rocket fuel residue on them, could be discounted because the absolute physical evidence—the parts of the Navy Standard missile—had already been secreted away under the cover of "national security."

CHAPTER THREE

THE AIR NATIONAL GUARD AND THE PRESS

The 106th Rescue Wing, New York Air National Guard, Westhampton Beach, Long Island, held a press conference on July 18, the day after the crash. Most of the pilots and pararescue personnel who were in the field when TWA Flight 800 was shot down were put before the cameras for one highly controlled display.

Since there was an Army unit further out on Long Island engaged in activity that included the launch of drones, a P-3 almost directly overhead to monitor infrared signatures of the military assets in flight during a military exercise, and because Linda Kabot took a photograph of a probable drone within this time frame, it is reasonable to believe the military would have personnel and equipment in place to play traffic cop during the initial stages of the drone's flight over Long Island and into the well-traveled waters to the south. Factor in the fact that Linda Kabot took her photo from a restaurant deck just east of the 106th New York Air National Guard airport and further, that the drone captured on the photo was flying toward the 106th New York Air National Guard base, and there is ample reason to believe the military units the 106th had in the air, and on the ground, were in place to monitor the initial stage of the drone's flight.

The 106th public affairs officer established several "ground rules" before the press conference began:

"We will not discuss the accident investigation since we do play a role in that, other than to make statements—these individuals have made statements to the FBI. What we will talk about tonight is what they saw and what they did."

Well, not quite. Severe ground rules were established prior to the press conference. No one would discuss the missile theory. No one would say they saw anything hit Flight 800. Nor would the helicopter pilots be allowed to say they were so close to the crash that they actually saw bodies falling from the sky.

The pilots had also been given a cover story to explain their presence in the area. They certainly would not be allowed to say they were playing traffic cop for a Navy drone that briefly flew over a civilian area before heading out to sea as the intended target of a Navy exercise.

The questioning began with Major Mike Weiss, pilot of a C-130: "We were out there doing a typical weekday/night training line. We were hoping to do some refueling with our helicopter guys and with a little bit of daylight left, we went over to our border training area just south of East Hampton, about ten miles off the coast, where we do rescue maneuvers, search patterns, testing of our equipment. And about that time, in the process of one of those maneuvers, our entire attention was directed to a flash of light to the southwest of our position, about fourteen miles or so. And we kind of rolled out and we could see a massive ball of flame.

"We just couldn't believe it, a huge pillaring tower of flame from maybe the surface to about 4000 feet. We were at about 1500 feet, and obviously it had caught our attention and we went straight over there. At the same time, our helicopters were on it. They had seen it west of the airport here."

An unseen female reporter asked: "Can you respond to other eyewitness reports that describe seeing an arc of light go from the bottom up?"

Major Weiss: "We did not see anything, arc of light or anything, prior to the massive explosions."

Major Meyer, the senior helicopter pilot in the air: "I was out on a routine training mission with Chris Baur and we were coming down the instrument approach, the ILS course to runway 2-4, which puts us northeast of the anybody field, facing to the southwest. At that particular moment, when we saw it, all three sets of eyes would look to the southwest.''

A female reporter, offscreen, cut in with: ''What about what you did see, can you describe it?'' But the 106th public affairs officer cut Major Meyer off, stating: ''Perhaps share what you might have said over the radio if it's printable.''

Major Meyer: ''What I saw, was, my first reaction was curiosity. I saw something that looked to me like a shooting star. Now you don't see a shooting star when the sun is up. It was still bright. It was probably just at the moment of sunset, and the sky was very bright, we hadn't reached nautical twilight yet. And I saw what appeared to be the sort of course and trajectory that you see when a shooting star enters the atmosphere. Almost immediately thereafter, I saw, in rapid succession, a small explosion and then a large explosion. And the large explosion engulfed the small explosion into a huge fireball that just then began to fall very slowly from the sky. And I was looking out at the ocean at a horizon. I can't estimate for you the altitude. It was certainly above me. I would say maybe, it appeared to be somewhere in the neighborhood of ten to twelve thousand feet.''

A male reporter offscreen asked: ''Were you concerned for your safety when you saw all of this?''

Major Meyer: ''No.''

''Why not?'' the journalist questioned.

Major Meyer: ''We were a long way away. We were

probably eight to ten miles off, well out of the trajectory of what we were viewing.''

Captain Chris Bauer also went before the microphones, but by this time the press seemed to be out of relevant questions. He was not asked about seeing a missile or a streak of light impact Flight 800. After the press conference, the FBI asked the Air National Guard to order the pilots not to talk to the press, according to an informant inside the Calverton investigation. Captain Baur is still not allowed to talk to the press. The official explanation is that his government employer, when he is not on duty with the New York Air National Guard, will not allow him to talk to the press. Major Meyer was eventually given permission to be interviewed by *Aviation Week*.

It was during this interview, closely monitored by a senior public affairs officer, that Captain Chris Baur said he saw an object going from east to west impact Flight 800. Sources inside the investigation had known this for some time but it had never been publicly revealed. And Linda Kabot says Major Meyer told her he thought she should publicly demand the FBI give back the negative they confiscated shortly after the photo was taken. Why did Major Meyer ask her to publicly take on the FBI? Because he thought the photo was of a drone.

CHAPTER FOUR

THE NEWS BREAKS

When the news of the Flight 800's crash into the Atlantic Ocean broke over CNN and the first pieces of film footage were broadcast, I was working with written notes I had brought with me on vacation for a book I was writing on POWs/MIAs. My wife and I were in California, 3000 miles away from the crash and visiting parents out in Riverside. Years earlier I had worked as a cop in Southern California, where I'd spent the bulk of my law enforcement career working traffic enforcement and accident investigation—the forensic stuff that goes into the reports to help determine the causes of an accident. After my medical retirement, I became an investigative journalist researching Defense Department and national security issues. That was what eventually led me to the stories about POWs and MIAs from World War II through Vietnam.

My wife, Liz, became a TWA flight attendant, and quickly moved up the ranks to in-flight training supervision. Although she worked out of New York and we lived in Virginia, she spent most of her in-service time at the TWA St. Louis hub and was gone about twenty days a month. On July 17, 1996, the day of the Flight 800 crash, we were vacationing in California and visiting old haunts.

I didn't know it as I watched television that night and learned about the crash, but immediately before the news release, the White House had received a preliminary FAA radar analysis indicating that TWA Flight 800 had collided with a missile. Even to professionals who were used to hearing shocking bits of reality, this was particularly gruesome. The intelligence services knew that there were very early reports from the FAA concerning a possible missile hit on the plane. It had turned up on radar. This story had to be killed immediately and a new story put into place.

It was about ten in the evening on the West Coast when I heard a newscaster characterize an earlier report of FAA radar tracking a missile on an intercept course with TWA Flight 800 as an "error." The broadcaster dismissed the FAA report not only as premature, but said that the government had reviewed the tape and now believed the blip that the radar operator had seen was an "anomaly." That was a pretty quick determination, I thought. The crash was only a few hours old, the rescue teams didn't even know what they were seeing out there in the water, but the FAA tape had already been analyzed to the point where the government could dismiss what someone had surely seen as an anomaly. I knew it was virtually impossible that a radar tape in New York could be adequately examined by experts using the latest technology to arrive at a final conclusion only a few hours after the crash. This kind of analysis, digitizing and copying and baseline-checking against other types of digital signatures, usually takes days, if not weeks. But this took place in only a few hours. Interesting. Why the rush to judgment?

My wife, Liz, was standing beside me as we heard the awful news. For as long as she could remember, she had wanted to be a flight attendant. At the age of forty, after our son had left the nest for college and I had matured to the point where I could be left alone for several days at a time without destroying the house, she began her flight attendant career with TWA. Now, more than ten years later, she was a flight service manager on semipermanent loan to the TWA training center in St. Louis. There was a demand for the kind of job Liz was able to do, that is, teach the new hires how to perform. For the first time in years, TWA was hiring flight attendants. Air travel was up and things looked promising at TWA. Every week at least one class of brand-new, eager young men and women were graduating and leaving for New York to work on "the line." And most of the graduates from TWA flight attendant school were Liz's former trainees and had become part of the TWA family. Any crash anywhere would have been a shock to her, because someone she knew or had trained would have been on that cabin crew. But Flight 800 was particularly devastating.

Knowing that she had close associates on the flight crew, she began crying when she heard the news, and with the tears still in her eyes, she began to return the phone calls that had been piling up all evening.

Some of Liz's new hires had been lost, shot down in flames, I would later determine. More than fifty fellow employees were on board. The fifty-four crew members—over twenty-three percent of the entire 230 passenger load—represented a higher percentage of fatalities than that of American combat personnel lost on the lines during the Gulf War. If it was truly an accident, what a tragedy. But if someone caused it, then it was murder. And I didn't like what I'd heard on the news. Maybe I'd been investigating Defense Department cover stories for too long.

Very early the next morning I jumped on the phone. Two federal law enforcement agents who worked out of D.C. said the inside rumor mill still insisted that the government had gone into a crisis mode because a missile had destroyed Flight 800. One source thought it was "friendly fire." The other said he hadn't heard who pulled the trigger. When government law enforcement people tell you that some agency inside the beltway has gone into crisis mode, it means that anybody who normally talks to the press doesn't, that grim-faced officials make sure that everybody tells the same story, and that they circle the wagons and take a defensive posture with respect to any question. They deny everything. The normal government procedure when no one is at fault is to become aggressive about finding answers and righteous about getting to the truth. When the opposite happens, you know someone is hiding something.

I began to call her TWA contacts in New York. "What really happened?" The rumors were almost endless. Two Air National Guard officers had confided that friendly fire brought down Flight 800. Crew members from an on-scene Coast Guard vessel had told TWA personnel the same thing. Had they actually seen anything? Had they only heard the story from others? I couldn't find out right away, but was truly disturbed that Coast Guard people would be talking about a missile one day after the government denied that FAA tapes showed a missile attack on the plane.

Several weeks after the crash, I walked into the TWA offices at the St. Louis training center. Liz was standing just outside her office as I approached. I could see that I was interrupting a very serious discussion that had been underway for a while. The small group of women turned on me as if I were the enemy. What was going on here? I was kind of a fixture around there, like an old chair. I tuned up their cars, took them out to dinner, listened to their problems, and dished out advice. Most people ignored me when I turned up to see Liz, so what was wrong?

I soon found out. Since the crash, some TWA flight attendants and other employees had informally collected information about what they were certain was a government cover-up in the making. No proof, just a theory. And I had agreed to pass the promising rumors they'd discovered on to my contacts in the press. But nothing came back with a solid ring of authenticity, despite the persistent stories of missiles on radars and mysterious photos showing streaks of light.

Anybody on the maintenance or engineering side at TWA knew what the NTSB was looking for in the hangar at Calverton. And even the experienced flight attendants who had no maintenance background knew what kept a plane up and what brought it down. Pilots knew. If you fly these things, you know what makes them tick. Therefore you also know when someone's lying. And that's what they said was going on at the NTSB investigation at the hangar. "This isn't just a mistake, Jim," members of Liz's group said. "Something's up big-time."

The NTSB "mechanical" theories leaked to NBC evening news on a regular basis were carefully orchestrated "tech talk" generated by unnamed federal sources. The NTSB crash investigators knew, not assumed, *knew* for a fact that the fuel transfer from the right to the left wing had begun more than four minutes *prior to* the explosion. They also *knew* as a matter of fact, not conjecture, that a spark-induced explosion *must* occur at the time the fuel switch is activated, not four minutes later.

Similarly, they knew that the scavenge pump story, the second NTSB theory put out by the federal government through releases to the networks, didn't make sense. There was a verbal procedure that would have been heard on the cockpit voice recorder announcing the commencement of the pump's operation. But it wasn't there. They knew that procedure was not on the tape, so it could not have happened had the cockpit crew been following normal procedure. Were they following normal procedure? The obvious place to look was at the scavenge pump switch itself, to see if anyone had turned it on. Again, no big secret here. The NTSB investigators already knew the scavenge pump switch was in the off position when they found it on the ocean floor. So why come out with a story that says an erratic spark or friction from an undermaintained or worn-out scavenge pump might have been the cause of the accident when they already had the scavenge pump switch and knew that nobody had turned it on? Yet here was the story on the nightly news. Someone from the federal government was deliberately feeding the national news organizations false information.

The people in the group that confronted me that afternoon at the TWA flight training center weren't radicals or extremists. They weren't liberals or conservatives. In fact, most of them weren't political at all. These were just dedicated working people who wanted not only to keep their jobs, but to keep the company they'd invested in alive and flying. If it had been an accident or a maintenance failure, they were ready to face the consequences, but if it hadn't been, and they and their company were being made the scapegoats for government blame and public wrath, they weren't going to put up with it. Their company and their careers were being sacrificed. They collectively held the notion that they lived in a country where the government could not do such things.

And that's what I told them—the government can't do these things because the Constitution prevents it. But the group in front of me wasn't interested in Constitutional theory. They wanted facts.

"But what if it really was a mechanical thing after all?" I asked.

"If it's mechanical, it's mechanical," they said. "But you have to find out. Because if it was a missile, we can't let them get away with it."

And they were right about one thing they said. If you have eyewitnesses to an event you have to completely discredit what they say before you can discredit the event. If only one witness holds up, for any reason, then the event must be given some credibility even if you can't explain it. The friendly missile theory seemed very remote to me at first for the very same reasons that the FBI kept on discounting it. You just can't torch off a missile, shoot down a plane, and not have people see the whole thing. But people did see it, a lot of people. In fact, there are just too many witnesses to streaks of light and not enough plausible explanations for them. On this basis alone, a guided missile shootdown would have to be at the top of the list.

I agreed to take a look at the federal investigation, such as it was. There were only a few TWA employees with regular access to the investigation. Unfortunately, they were not involved at a high enough level to have direct access to the hard evidence. Therefore, it would be necessary to get inside the FBI or NTSB investigation. This would be a difficult task.

I also knew from my law enforcement days that the FBI has great technical expertise and even better press relations. Separating FBI public affairs statements from the truth could be difficult, especially in this case. Intermingled with FBI public affairs were the NTSB, White House, and Pentagon public affairs bureaus, each feeding the media its own version. As the stories get more repetitious and the public loses interest in the crash, it's pushed off the front pages by fresher news, and the chances that the government will be able to bury the story for a few more years become greater and greater. Maybe lawyers for victims' families can win a settlement from TWA's insurance company. Maybe enough real evidence can be suppressed and enough people frightened into silence for the government to let the story die away.

Once in a while a reporter or two might pop up to review the case and ask what's happened in the intervening years. But he'll be told that there never was evidence of a criminal act so the FBI terminated its involvement. The NTSB never really found the cause, so there isn't much that can be done.

I knew from my Defense Department experience that the chances of inside-the-beltway "reality" being anywhere near the truth were usually very remote. The press was eating out of the hand of public affairs offices, and, with no information of their own to go on, were taking whatever they could get from the official sources. But we have a very big government with lots of different people inside it. And with rare exceptions, the official government line is only one side of the story, even within the government. For example, some of my D.C.–area law enforcement sources heard there was considerable frustration inside the NTSB and FBI investigations, especially at the foot-soldier level.

In a real cop shop, I knew from experience, there is absolutely no chance of a high-level cover-up succeeding if street-level personnel were involved in the case. Street cops traditionally dislike management. But this wasn't a real cop shop situation. These were accountants and lawyers with badges who did what they were told. So in order to get to the truth, I'd have to go in through the bottom to sniff out the discontent and find out who was beginning to grouse about the truth's getting shafted. That's where I'd find the beginning of the case.

So I told the group of TWA personnel at the training center in St. Louis that I'd sniff around, do what I could, and if I could find evidence to support our suspicion that a missile had brought down Flight 800, I'd get it out there for everyone to hear.

CHAPTER FIVE

INSIDE THE CALVERTON INVESTIGATION

"If you had called me even a week ago, I wouldn't have spoken to you about Flight 800," the caller said over the phone. There was a pause and I could hear him breathing. This obviously wasn't easy for him and I wasn't going to fill up the silence at his end of the phone with words.

"What's going on here . . ." he said, "what's going on here—it's just not right. We're supposed to be conducting an investigation, but we're not. It's just busy work and theories with nothing to back it up."

The source went on to say that he had decided to talk to me because the evidence of a deliberate cover-up was so overwhelming inside the investigation that he wasn't willing to continue to ignore its implications. Nor, he said, was he willing to put his job or career on the line. But he was willing to talk.

I told him I was a writer and that eventually whatever he said to me that could be verified, substantiated by evidence, would find its way into a story I was writing. He understood, and we began to talk.

Several weeks later, we agreed to meet at an old hotel near the south shore, a few miles from the Calverton hangar, where the NTSB had brought the airline wreckage for reassembly and evaluation. It was where the crash investigation field headquarters was located, and where my source told me the actual cover-up of information was underway.

At our first meeting, my source, explained the process of the recording of the wreckage and other pieces of evidence in the crash. What he eventually described was actually a noninvestigation more than a search for the truth.

Once the Flight 800 debris got to Calverton hangar, he said, the NTSB teams on the floor did a credible enough job of tagging and logging in the evidence. But there was no actual NTSB investigation. The Fire and Explosion team, headed by Dr. Merritt Birky, controlled NTSB "mechanical" theories, the source told me. Birky's crew would seize upon a piece of debris, such as a center wing tank (cwt) fuel pump. They would examine it for any signs of obvious damage, any telltale signs of stress or wear that in and of themselves might have led to the kind of failure that affects other components in the system, and whose failure can lead to a mechanical catastrophe. If they found any suggestion that the part might have failed, even in the smallest way, they would theorize about how this part could have caused the plane to crash. But it was all speculation. The NTSB team did not try to back up any theory by analyzing the debris for a chain of evidence.

Nor did anyone on the NTSB team assemble or even sketch out in a rudimentary fashion a meaningful diagram of the debris itself which had already been accurately located by a satellite global positioning system (GPS) used by the divers to develop a map of the debris field correct to within a couple of feet. However, the NTSB had built a computerized depiction of the 747 at the top and the bottom of the ocean with hundreds, perhaps thousands, of lines descending from the plane to the ocean. So, for example, if the R/A gear box for fixed wheel drive were found at a certain spot on the ocean floor, there was a line connecting it to where it would have been were the plane sitting in one piece on the top of the ocean. Somehow, the NTSB believed that this type of graphic layout would help show them how the plane had broken up.

But according to my source, it was actually a useless mess. Sure it impressed the press and public each time it was used as a prop during the periodic FBI/NTSB press conferences on Long Island, but as an investigative tool it held very little value.

I also believed, as I listened to the source describe

what was going on in the hangar, that the NTSB was adept at substituting excellent public affairs and politically motivated theories and "findings" for basic investigative competence. They leaked bits and pieces of what they knew through unidentified sources or background sources who could not be named, to move the public perception of the investigation along to a new dot on the connect-the-dots diagram of what they wanted people to believe was the truth. It was a simple enough procedure. The news organizations must have thought that no competent senior level NTSB investigator would cause a theory to be disseminated nationwide unless there was substantial evidence to back it up. My source, however, insisted there were absolutely no facts to back up the reports being leaked to the national television press.

"In other words, Jim," he said, "whatever you're hearing in the news, whatever Tom Brokaw says, comes from someone at NTSB who says he's releasing the latest bit of information from inside the investigation. There's not a shred of evidence to back it up, because it's only conjecture. In fact, the evidence is going in the opposite direction."

Months later, an NTSB document would be handed to me proving that no evidence had ever been found of a mechanical cause for the loss of Flight 800.

The source seemed most disturbed by this steady drumbeat of high-level, unnamed government leaks painting a false picture of a mechanical failure when, in fact, there was no evidence to back this theory.

But there was significant evidence pointing toward a missile. More than 150 people reported seeing what they said was a missile intercept Flight 800. Thirty-four were found to be credible by the FBI. Each of those found credible were taken to the exact spot where the missile was observed. Surveyors set up their equipment at the exact spot and, using the credible witness to guide them, established precisely where the missile was first observed and its flight path. Every credible witness described a missile shot originating in military zone W-105, south-southeast of Long Island. And every credible witness described a missile shot that intercepted TWA Flight 800's flight path, according to what my source told me he had heard from the government investigators themselves inside the hangar. My sorce emphasized that no one inside the investigation had *concluded* that a missile brought down Flight 800. It was, however, the only theory with a viable base of evidence.

The orchestrated debunking of a viable theory by anonymous senior-level government officials, combined with what was going on inside the investigation, had led my source to suspect he himself was part of a politically correct accident investigation. In other words, one that could not reach certain conclusions regardless of the facts because the decision had already been made about which conclusions to avoid.

"And the FBI can come and take whatever they want from right under our noses," he said. "It doesn't matter where the piece comes from. If the FBI wants it, it's gone."

My source described how FBI personnel could enter any part of the Calverton hangar and remove a piece of debris without logging it out. There was no way to know if it was ever brought back without conducting an inventory costing millions of dollars. These pieces of debris would be taken to the FBI's on-premises "bomb room" at which point it simply disappeared from the NTSB's side of the investigation. I was told it was not possible to track any given piece after this point unless it was physically observed in the bomb room.

A few weeks after our first conversation, we met again at a hotel room in the middle of Long Island, close enough to Calverton to allow him to get back and forth without anyone noticing his movements.

We were waiting to begin our conversation when there was a soft knock at the hotel room door. My source actually jumped. He was so nervous at meeting me at the very hotel where many of the Flight 800 investigators stayed that even the slightest indication that someone was outside in the hall stopped his conversation in mid sentence. But the documents he had were too important to leave to chance or to FedEx. As the informant placed his briefcase on the table and began to remove a large folder, someone loudly knocked on the door. The source turned pale and began looking for somewhere to hide.

I'd picked a bad time to order room service, I guess.

THE DOWNING OF TWA FLIGHT 800

Moments after the waiter left, my source was also gone, disappearing into the gray darkness of the corridor and the anonymity he carefully guarded. I began to read through the package of documents left on the table. Every piece of Flight 800 wreckage found on the ocean floor was precisely described and identified by longitude and latitude. Boring stuff to most people, but I'd spent many years in law enforcement as an accident investigator, and knew these documents, would tell the story of Flight 800's demise, once they were put into a computer and analyzed.

Soon I would know exactly what the FBI and NTSB investigations knew. Was it really a puzzle too complex to solve, with no clues leading to an answer, as government officials maintained at their weekly press conferences? I decided to break the data into two separate forms, a schematic of the Flight 800 cabin interior and a computerized mapping of the debris found in what investigators labeled the red, yellow, and green zones south of Long Island.

I started with the cabin schematic, the order in which the seats, victims, and galleys exited the plane. The first structural part to fall off the 747 was from the right wing, where it attached to the fuselage. Then the R-2 door came off. Next were four seats: Row 19, seats 4, 5, 6, and 7, directly inboard from the front edge of the right wing.

As I laid it all out, I could watch the pattern as the damage continued to march in a straight, narrow line across the cabin: Row 19, seats 8, 9, and 10. Then came the "C" galley, which takes the space where row 17, seats 4 through 7 would have been. Rows 18 and 20, seats 4 through 7 came next, followed closely by a small chunk of the front of the left wing, approximately where it attaches to the fuselage. And finally, the march across the fuselage was complete with the exit of row 19, seats 1 through 3.

At this point, a strong possibility existed that something impacted the 747 in the space between the R-2 door and the front edge of the right wing. Rows 17 and 18 occupy the center of this space. The damage continued along a very narrow line across the fuselage, appearing to exit on the left side of the fuselage somewhere between rows 18 and 19. In other words, something had impacted from the right, the south, or ocean side of the fuselage, continued moving at a destructive velocity across the cabin without exploding and exited the aircraft on the left side—toward Long Island.

The damage was almost fifty feet forward of the center wing tank fuel and scavenge pumps, where the NTSB for months had theorized the explosion originated.

I decided to begin the tedious process of inputting the data to create a computerized map of the 747's debris trail. Using a spread sheet to catalog the location of each piece, I hoped that by graphing it, I could get an investigator's bird's-eye view of where the debris laid out on the ocean floor and at what point during the time line of the breakup of the plane the debris came to rest in the ocean. I was actually working with three coordinates: latitude, longitude, and a chronological time line from takeoff to the ocean floor. By the time I was most of the way through the computerization, the implications of the cabin schematic unfolding before me were so enormous that I wanted to "see" the breakup of the plane from all possible views. But what I had seen to that point, combined with corroborating description from credible eyewitnesses, suggested that a missile with an inert warhead traveled through the plane. The warhead had to have been inert because there was no debris pattern indicating an explosion at the point of impact. Who would have used a dummy warhead? Certainly not the Hezbollah.

CHAPTER SIX

RECONSTRUCTING THE EVENTS

The almost five-foot-long, color-coded, computer-generated debris map, constructed directly from FBI/NTSB documents, visualizes a more-than-13,000-foot trail of debris and three distinct phases of the plane's loss:

—The initiating event.
—The center wing tank explosion.
—The right wing tank explosion.

Based on this map and information from sources close to the investigation, relating to the forensic analysis of the debris and victims, I was able to re-create a precise chronology of the crash of Flight 800.

Something impacted the right side of the 747's fuselage, aft of the R-2 door and in front of the right wing. The damage marched across the fuselage, exiting the plane just in front of the left wing. A chunk from the front of the center wing tank broke off and the front of the tank was weakened by the energy of the force passing through the fuselage several feet in front.

One of the two potable water tanks directly in front of the center wing tank was destroyed, the second only modestly damaged. Air conditioners and other portions of the 747 in the area at the point of impact began to peel away into the 400-mph wind.

This phase consumed almost eight seconds of flying time and 4700 feet of flight path. A phenomenon occurred during these 4700 feet that further suggests a missile: virtually all cabin debris exited through the hole in the left side of the aircraft, leaving a trail two- to four-tenths of a mile to the left of the debris that fell from the exterior of the plane. So, whatever entered the right side of the plane exited the left side, leaving a pressure imbalance that sucked passengers and hardware out of the cabin through the left side hole.

About eight seconds in to the breakup of Flight 800, there appears to have been a "mild" explosion in the center wing tank. "The pattern of explosive damage in the tank loosely resembles a triangle, starting at a point in the center of the tank and expanding toward the front of the tank," a senior investigator said. This modest explosion bowed the top of the center wing tank upward, also bowing up the aluminum cabin floor directly above the cwt. But the explosion did not penetrate the roof of the cwt. Instead, it blew out the weakened front of the tank. This, in turn, completed the separation of the fuselage in front of the wings. It tumbled end over end, slightly to the left of the plane's track, rapidly falling away from the remainder of the plane, which continued on.

The center wing tank was not a factor in what caused the plane to crash. It is relevant only to a second event, about eight seconds in to the plane's breakup.

For the next 5000 feet the remainder of the plane, from the front of the wing to the tail, continued forward in a dive. The fuselage and wings, which were still attached, gradually rolled to the left until the left wing pointed toward the water. The fire from the central wing tank traveled up, into the right wing vapor vent and along the fuselage closest to the right wing. Even the most inexperienced accident investigator, looking at the path of the flame from the cwt to the right wing vapor vent could see that the flames didn't travel laterally. Heat rises, so the path of the flames should have indicated that the left wing was *below* the right wing when the flames began to move.

About 10,000 feet after the initiating event, an explosion blew the number one and number two engines attached to the left wing downward along with the majority of the cwt. The center of this debris field came to rest on the ocean floor almost 12,000 feet east of the initiating event.

The number three and number four engines were lofted upward by the right wing tank blast. The 8500-pound engines landed in a debris field about 1000 feet east of the number one and number two engines. The remainder of the cwt also landed in this debris field, almost 13,000 feet east of the initiating event.

When properly laid out in a diagram, the debris field was not difficult to analyze. Each phase of the breakup could be clearly seen. This was not a crash scene of such complexity that hundreds of investigators could not find an initiating event.

At this point, the only theory that comfortably fit the evidence was a missile. Bomb damage would have radiated out from a central point, creating a distinctive signature on surrounding portions of the plane. This did not happen. Nor would a shaped charge leave an initially narrow debris path across the entire width of such a large aircraft, at least not without also leaving the signature of an explosion as the initiating event. This, too, did not happen.

A catastrophic failure of the fuselage caused by metal fatigue would have resulted in the immediate disintegration of the aircraft. Evidence of metal fatigue would have been found in the Red Zone, closest to JFK airport. But the plane did not immediately disintegrate and, according to the NTSB's own document: "No evidence was found of fatigue in [the] fuselage structure for pieces associated with the "red zone"." Absolutely nothing associated with mechanical failure can be linked to any event or series of events occurring during the first eight seconds after the initiating event.

These facts come from the FBI/NTSB's own documents and were known quite early in the investigation.

My analysis of the initial documents was complete, and it was time for me to refocus on my best source. During the third week in November, we met in Washington, D.C., this time away from a crowded hotel and room service. I wanted to lay out my best material in front of the one person who had an eyeball view of the entire wreckage assembly in the hangar. He would know almost immediately whether my analysis made any sense when compared to the actual debris that had been recovered and systematically reassembled on the hangar floor. It was a moment of truth, both for me personally and for my ideas about the crash.

When I showed the source my 747 schematic showing something entering aft of the R-2 door, traveling through the fuselage, and exiting just in front of the left wing, the source quietly whistled and said he had just seen the same analysis inside the investigation.

"What do you mean, 'seen'?" I asked.

"Seen," he repeated. The NTSB had already seen and described the same pattern I had put before him. Now the NTSB Sequencing Team was perplexed. It was not possible to fit the facts of this phantom object ripping through the interior of the 747 with any

mechanical theory they could come up with. But on
the NTSB side of the investigation, the mechanical
theory was all that would be considered. If it became
impossible to publicly defend "mechanical," the
NTSB would be out and the FBI would become the
lead investigating agency.

According to the source I was interviewing, the
Suffolk County coroner had informed the NTSB that
he would not go along with any of their theories about
the cwt being the primary cause of the crash. The
coroner had exhaustively inspected the reconstructed
cabin seating and victim damage, computerizing and
correlating the damage, and it did not fit the NTSB
cwt theory.

The source also mentioned in passing that there
was no victim burn damage at all. Because of the
absolute nature of the statement, I asked him for
clarification. He said it was his belief, based on information gathered while at the Calverton hangar, that
significant victim burn damage had not occurred, at
least not while the victims were still in the plane.
There were isolated cases of burns, perhaps, but no
pattern leading to any conclusion. (I later received
documents revealing that almost all the bodies with
thermal or chemical burns were among the eighty-
seven recovered floating on the surface, which had
been a mass of flames for hours after the crash. So
these bodies were consistent not with an explosion
in the cabin but with postmortem burns that took
place after the bodies had hit the water.)

The source sat there quietly looking at a cabin schematic of the 747-100. He turned to me and said, "If there was missile exhaust residue anywhere inside the cabin, where do you think it would be?"

I pointed to a general area in front of the wings and drew a line across the cabin to a point just in front of the left wing. The source commented, "You're right on the money."

I sat there stunned. "Are you telling me there is missile exhaust residue inside the cabin mock-up at Calverton?" I asked.

The source said there was a reddish-orange residue on the back of seats in the area in front of the wings.

"Is there residue anywhere else?" I asked.

"Not to my knowledge," was the reply.

I asked my source to closely inspect each seat on his next trip to Calverton, and write down all seat numbers with the reddish-orange residue. The source agreed to do so.

"Does anyone else know there is reddish-orange residue?" I asked.

"Oh, yes. The FBI has taken samples," was the reply.

"And?"

The source said he had not seen the lab results, but the missile theory had suddenly become very credible to the FBI teams in New York. NTSB investigators working at Calverton had repeatedly asked for the results, but the FBI had refused to cooperate. There was even talk of James Kallstrom paying a visit to the

China Lake missile testing facility in California to witness a shot by a missile with an inert or dummy warhead.

Reddish-orange residue along the exact path revealed through analysis of facts gleaned from the debris field presented me with the probability that solid evidence existed within the Calverton hangar that a missile brought down Flight 800.

According to the NTSB debris tag report, concrete forensic evidence of a missile passing through the fuselage of Flight 800 was available to the investigators as early as August 3, 1996, when seats 1, 2, and 3 from row 19 were recovered. Each seat, easily verifiable through an eyeball inspection of the seat fabric, had a significant amount of a reddish-orange residue along the back. Many of those who witnessed a missile rising from the ocean and intercepting Flight 800 described a "reddish-orange" flarelike object. So what happened to the investigation into the reddish-orange residue?

The weight of evidence was strong enough without the residue to conclude that a missile was the culprit, unless of course the government had a vested interest in keeping that information from the public.

If it is missile residue, the U.S. Government, by its actions, is in some manner responsible, and the FBI, at the senior levels, knows it. And, if the FBI knows, we can only assume that Justice, the White House, and the Pentagon also know, because the FBI is just the investigating agency. Political responsibility for

FBI actions in such a sensitive matter lies at the very top of the executive branch.

If a foreign government or terrorist organization were responsible, the FBI would have taken over as the lead investigating agency. This has not happened.

A cover-up of unprecedented proportion appeared to be unfolding before my eyes with every new piece of information.

CHAPTER SEVEN

MILITARY DOUBLE-TALK

At this stage of the investigation, I teamed up with David E. Hendrix, a longtime reporter for *The Riverside Press-Enterprise,* and the first journalist to come up with hard evidence the government was hiding something. Considering that Hendrix's desk is almost 3000 miles from where Flight 800 crashed, it was an extraordinary journalistic feat.

Within hours of Flight 800's going down, a long-time Navy source of Hendrix's passed on Navy scuttlebutt that it had been brought down by friendly fire. "An ex–Navy officer, who used to supervise warning areas and spoke on the condition his name not be used, said he was told the plane was the victim of an exercise gone awry, a practice mission involving the

Coast Guard, U.S. Customs, and the Air National Guard,'' read Hendrix's article.

Hendrix didn't know if his source's information was accurate, but he had been reliable in the past. If the source had one fault, it was giving analysis along with hard facts and forgetting to separate the two.

Hendrix decided to dig deeper and immediately ran into a succession of government public affairs officials. But Hendrix is an unusual reporter. He recognizes how the federal bureaucracy operates. If you displease the political ruling class too much, they will not just counter your journalistic efforts, they will go the extra distance to destroy your reputation.

Few venture into this minefield. Most that do, scurry for cover when the bureaucracy growls. Like my initial reaction to the government's denial of the missile story in the early hours after the crash, Hendrix's own internal alarm had begun blaring. So he continued digging. Another Navy source, on the condition of anonymity, confessed that a submarine was involved in an exercise in military zone W-105, southeast of the Flight 800 crash site. At least one Navy P-3 antisubmarine warfare plane out of Brunswick Naval Air Station in Maine, was also involved in the exercise, the source said.

Later, the Navy would also admit that the AEGIS cruiser U.S.S. *Normandy* was in the area, but 185 miles south of the Flight 800 crash. For good measure, the Navy also volunteered that the *Normandy*'s radar was not operating at its longest range. In other words, the *Normandy* could not see Flight 800 but that was disingenuous because in a CEC exercise a warship like the U.S.S. *Normandy* that far over the horizon was not supposed to "see" the flight. It was supposed to react to radar data fed to it by the units over the target, like the Navy P-3 Orion.

Hendrix began to work his way through the maze of Navy public affairs officers, looking for answers. He received several documents from a source within the Federal Aviation Administration. Priority message R015060 was sent from the Naval Air Base, Oceana, Virginia, to the FAA, Ronkonkoma, Long Island at 11:07 P.M., July 16, 1996, advising the FAA of a phone conversation between a Ms. Cosby at Oceana Naval Air Base and Mr. Dombrowski of the FAA in which a zone south of Long Island was reserved for military use as of 8 P.M., July 17, 1996. The FAA facility at Ronkonkama is about fifteen miles north of where TWA Flight 800 would crash less than twenty-four hours later.

But the FBI and Navy public affairs officials continued the denials that any significant amount of military activity was scheduled the night of July 17, until Hendrix faxed the Navy copies of the FAA documents, then the "Navy representatives in Washington, D.C., and Virginia ended discussions and referred all questions to the FBI and NTSB," according to Hendrix.

The bureaucracy was caught lying about vital facts related to journalistic investigation of the death of 230 people. A reasonable person would expect the press corps to rally around the exposure of false government statements about a military warning zone being reserved for military activity thirty-one minutes before TWA Flight 800 went down. But the general press corps was silent.

The Navy controls the offshore training areas south and east of Long Island and notifies the FAA whenever one or more are to be restricted for military use. So this was nothing more than the bureaucratic two-step in the face of incriminating evidence. Lie until you are caught, then pass the buck to the next agency along the line, one not only better suited to handle the information but, because it was out of the loop when the first false response to questions came down, it had more credibility than the agency which had issued the denial. It was a game that government agencies, notoriously careful about disclosing any facts to anyone, are very adept at playing, and the military, NTSB, and FBI are old hands.

Nevertheless, Hendrix continued to dial the phone. The NTSB refused to return his calls and FBI spokesman Valiquetie said his agency would not comment either, especially about whether its investigators knew about the military exercises in the warning zones off Long Island. He specifically declined to examine copies of an FAA log, map, and Navy message documenting the activities in and around the warning zones on the days leading up to and the night of July 17, 1996.

But Dave assured me that the *Press-Enterprise* was

going forward with its investigation and that they were continuing to develop their own leads. We would ultimately work together down the line, but Dave believed he smelled the same rat that I did and was off on the chase to develop whatever information he could.

Now, at least, a part of the story would see the light of day. What I didn't know was just how aggressively hostile the government would become when shown the evidence of its own cover-up.

Just why were they trying to hide the truth with such vehemence? Was there a bigger story here that I was missing?

CHAPTER EIGHT

POLITICAL SHELL GAMES

A few weeks after the Hendrix article ran in *The Press-Enterprise,* he received a late-night phone call from Washington, D.C.—very hush-hush. Political operatives who claimed to be independent professed to have a great yearning to get at the truth: What really brought down TWA Flight 800? They offered to pay Hendrix's airfare if he would get on a plane within forty-eight hours. Drop everything, they said. This is important. We want to get to the bottom of this as much as you do. Who knows, perhaps we will be able to help you find something out.

Hendrix demanded assurances that this was a legitimate congressional investigation and not a last-minute desperation attempt to embarrass the Clinton Administration. He was given a verbal assurance. Congressmen and other officials were stuck. What questions should they ask? Hendrix demanded that the flow of information go both ways. He had questions that could easily be answered by congressional investigators attempting to get at the truth.

You tell us what you want to find out, they assured him. Send us your list. So Hendrix drew up a two-page list of questions, duplicates of those he had asked the Navy, FAA, NTSB, FBI, Coast Guard, and others and faxed them off:

1. Who or which command agency or unit individually asked to reserve Warning Areas W-105, W-108, R-4001, and the special area requested by 16 July 96 message from FACSFAC?

2. Were there any other special exercise areas requested or activated off the East Coast for 17 July 1996?

3. Who was/were the requesting officers, officials, or agents?

4. Have the NTSB, FBI, or any other investigating agency talked to them?

5. Why were the areas requested? What kind of specific exercise, activity, or project was underway?

6. What units, organizations, ships, aircraft, satellites, or other intelligence or data-gathering assets were involved for the requested activities?

7. What units, organizations, ships, aircraft,

satellites, etc. ultimately participated in the operations?

8. Did any units drop out? Which ones? Why?

9. Where are the messages, reports, requests, memos, after-action reports, accident reports, exercise summaries, evaluations, and other documents relating to the use or scheduled use for the designated warning areas or special areas? Are there any messages, computer printouts, e-mail, accident reports, distress calls, reports, summaries or any other (documents) relating to an accident involving TWA 800?

10. Were the areas used for multiple tasks, i.e. for Coast Guard, Customs, Drug Enforcement Agency, National Guard, Navy, etc. that ran concurrently or in succession?

11. Can other agencies, i.e. Coast Guard, state police, Border Patrol, etc. use the areas without alerting the Navy or having prior authorization?

12. What munitions, including practice or inert rounds or missiles, were used in the exercises? Who did the evaluating?

13. What was going on or scheduled to go on between the surface and 6000 feet in Warning Area W-105? Even if these areas are routine exercises, what happened? Define routine: something that happens frequently or something that is a dangerous activity? Did any munitions, including practice or inert rounds or missiles, go astray?

14. What was going on or scheduled to go on between the surface and 10,000 feet in the specially requested area identified in Navy message R015060 of 16 July 1996? Were any munitions, ordinance, missiles, etc. lofted between the special exercise area(s) and the regular warning tracts?

15. What was going on or scheduled to go on between the surface and 11,000 feet in Warning Area W-108?

16. What was going on or scheduled to go on between the surface and 10,000 feet in Warning Area R-4001?

17. What was scheduled for Warning Area 386 between the surface and 24,000 feet as of 12:01 A.M., 18 July 1996? Did the exercise occur?

18. Did the requesting/approved agencies have other units, agencies, groups with them, as auxiliary or tangential units, that could have fired any munitions, including practice or inert rounds?

19. How often are munitions, ordinance, missiles, etc., including practice or inert rounds, fired or used in the warning areas or special practice areas? How close are such munitions, missiles, etc. permitted to come to unrestricted air space?

20. How many P-3s were in the air along the East Coast the night of July 17, 1996? Why were they aloft? What were they doing? Were they

involved in exercises with other organizations? If so, which ones? What was their mission?

21. Were any of the P-3s involved in any exercises or activities measuring the accuracy of missile or other trajectory munitions or electronic simulation of such firing?

22. Was the Air National Guard or any other unit involved in exercises that involved dropping flares, chaff, or other objects for use as targets or EW [electronic warfare] programs to disrupt practice missiles or simulated missile exercises?

23. What U.S. surface ships, including but not limited to Navy, Coast Guard, Army, were active offshore? What were they doing? Were any crews or personnel from other agencies aboard who had missiles, shoulder or otherwise, live or inert or practice? Were any missiles fired? Did they strike TWA 800?

24. What was the name of the submarine involved in the joint exercise that took place the night of 17 and 18 July, 1996? Was it an American sub? If not, from what navy? What was it doing? Who was aboard? Were there any other military units aboard, such as [Navy] Seals, Special Forces, etc.?

25. Were there any other submarines active the night of 17 or 18 July, in or out of exercises, underway to somewhere else or just cruising?

26. Was the U.S. military or any other U.S. agency conducting exercises with any foreign

interests or agencies? If so, with whom? What was the nature of the exercises?

27. Do any American submarines, nuclear or otherwise, have antiaircraft missiles or capabilities? do any, as a matter of course, store shoulder-launched missiles aboard for possible use by landing parties or Seal units?

28. Is the U.S. Navy experimenting with anti-aircraft capabilities aboard its submarines? Is it experimenting with technology acquired from the ex–Soviet Navy or its former republics?

29. Has the United States bought, traded, purchased, captured, or acquired any foreign vessels, submarine or otherwise, that have anti-aircraft missiles or are outfitted with antiaircraft missiles or projectiles and are being used, stud-ied, or practiced with and were active the night of 17 July?

30. What satellite imagery is available? What do satellite photos show about the preexplosion and explosion of TWA 800? Can we have copies of the images/photos?

31. Have U.S. officials asked any foreign gov-ernments for satellite imagery they might have, showing missiles or any other foreign object, pertaining to TWA 800? What did they have? What was shown? Can we have copies of the images, photos?

32. Did the NSA or NRO pick up any electronic intelligence regarding TWA 800? Has anyone checked with the NSA or NRO?

33. What does Israeli intelligence say about the cause of TWA's demise? France? Great Britain? Russia? Iran?

34. What happened to the original radar tapes that showed blips or anomalies near TWA 800? Has a laboratory independent of the FBI or other U.S. agencies examined the tapes?

35. May we see copies of TWA 800–related e-mail among commanders and subordinates in the first forty-eight hours after the crash?

36. May we see copies of orders or memos to divers looking for TWA debris?

37. Where do the Navy diving teams come from? Which units? Who sifts the debris to determine what is going where? Did the FBI/NTSB know the warning areas were activated?

Hendrix was assured that all his questions would be answered in D.C.

When he landed at National Airport he was met by Dole presidential campaign operatives and the wife of a very high ranking Republican. It was a very difficult position.

This wasn't a group of investigators looking for additional information to buttress a case under development. It was a group of political hacks with no information except what they had read in the newspaper. In meetings held over a two-day period, it became obvious they either had no desire or no real support to investigate TWA Flight 800 independent of political concerns.

CHAPTER NINE

CLASSIFIED: THE NTSB INVESTIGATION REPORT

In December 1996, an informant handed me a classified ten-page report entitled "TWA 800, Chairman's Briefing/Status Report, November 15, 1996." It was not on letterhead, nor was it signed. The informant said the "Chairman" was James Hall, who headed the NTSB. Rumor had it that *The Washington Post* already had the document but could not confirm it.

If this actually was the NTSB chairman's report, it was a very hot property. In ten pages the chairman exposed a serious conflict between the FBI and NTSB investigations. The document also revealed that on the night Flight 800 crashed, an FAA technician analyzed the radar tape and concluded that a missile was seen on the radar screen on a collision course with

Flight 800. This information was not immediately given to the NTSB as federal guidelines mandated. Instead, the information was forwarded to the White House and the tape was sent to the FAA Technical Center in Washington, D.C.

The most damaging fact in the documents was the revelation that the NTSB had no evidence to back up its often-leaked theories about the center wing tank exploding due to static or other fuel-transfer problems. If this report was legitimate, Tom Brokaw and the *NBC Evening News* team were being taken for a ride by the NTSB, because they had become the almost official "leakee" of choice whenever the NTSB felt the need to inject the mechanical theory into the minds of America's citizens.

But how to confirm a document one of America's leading newspapers had allegedly struck out on? I had coauthored two books with a journalist named Mark Sauter when he was an up-and-coming young television reporter at KIRO TV in Seattle. Sauter is a unique individual. He graduated from Harvard, then became a Special Forces officer before returning to civilian life as a graduate student at the Columbia University Graduate School of Journalism. The aggressive Special Forces attitude never left him. When on the trail of a story, he attacked.

Mark Sauter's job title is Investigative Correspondent, *Inside Edition*. Over the years he has worked both print and television news. I had occasionally passed Flight 800 tips to Sauter in the months after the crash, but none had resulted in stories. Apparently, though, they had raised enough curiosity that *Inside Edition* asked me to be "their eyes and ears" for Flight 800. *Inside Edition,* however, leaned strongly toward the NTSB mechanical theory in spite of the complete lack of documentation or physical evidence. The rationale seemed to be that TWA had not aggressively defended itself in the weeks and months after Flight 800 crashed. It had not established contact with trusted reporters and fed them the nonmechanical side of the story. The unrelenting flow of mechanical theory leaks by the NTSB, unchallenged by TWA, constituted prima facie proof of guilt for many journalists.

Now Sauter was in New York City. It wasn't the hard news he was used to on the evening news in Seattle. But he wasn't tied to a local area. So I flew to New York and showed the ten page NTSB memo to him on a cold December day, less than two weeks before Christmas. The following Monday, Sauter attacked. He spent twenty-four hours dancing around with the NTSB public affairs personnel in Washington, D.C. They were using the typical public affairs techniques, "trying to damage-limit whatever they thought would come out of that memo."

According to Mark, the representatives of the federal agency alleged to be hard at work getting at the truth acted suspiciously like they had something to hide. They tried not to say anything on the record or on background. Sauter had faxed the NTSB selected parts of the memo as part of his strategy. They admitted off the record that the document was real, at least the parts Sauter had faxed. The NTSB public affairs officer said, ''there were errors'' in the document. But he would not say which specific portions of the NTSB memo were incorrect. It was their memo. When written, every word was alleged to be true. Now that they were in a damage-limiting mode, they wanted to imply that nothing in the memo was true.

After a day of maneuvering, the NTSB said one thing wrong with the memo was a reference to the FBI's not returning photos to the NTSB. It was a minor point. But the NTSB left the issue of errors hanging by suggesting this was only one of multiple errors. During this verbal sparring, the NTSB let slip that their chairman, James Hall, was testifying on Capitol Hill the next day, but they refused to say where.

Normally this type of information is readily available. Perhaps the chairman was hiding from *The Washington Post,* alleged to be trying to confirm the NTSB memo written by Hall. Whatever the reason, his upcoming testimony was not part of the public record. But Sauter learned that the chairman was testifying on air bags in front of a congressional committee. The next morning, Sauter was on a plane headed for D.C. When he arrived at the hearing on Capitol Hill, Sauter sent the film crew into the crowded hearing room for the typical talking head shots seen on the evening news.

Sauter warned the public affairs staff hovering around the chairman that he and the film crew would be waiting just outside the door, the only exit. The public affairs person was told that Sauter was going to talk to the chairman about the inability of NTSB public affairs personnel to answer very simple questions about the accuracy of the chairman's own memo. While Sauter and film crew waited in ambush, an official came out to greet them. "Are you waiting for the chairman?" he asked.

"Yes," Sauter responded.

"He's not going to answer any questions," the man said.

"We'll see," was Sauter's response.

"But, but, maybe you'll talk to me, I'm Peter Goelz, the head of governmental and media relations for the NTSB. Maybe I'll answer some."

"On camera?" Sauter questioned.

"Uh, I guess," was the less than self-assured response.

The camera began to roll. Peter Goelz was Director of Public and Government Affairs at the National Transportation and Safety Board.

"What is this report we're talking about?" Sauter asked.

Goelz replied that it was a draft of "working minutes from a regularly scheduled review of our investigation."

This statement confirmed the report in its entirety. Unless Goelz wanted to confess that everything the

117

chairman wrote was incorrect, the ambiguous suggestion that Sauter possessed a false document was disproved by the number one NTSB public affairs official.

"We hold these kinds of meetings periodically, where we just get caught up on what everybody is doing. We all try to sit down in one room and review where we are. And that was the first draft. From what I can see, it certainly wasn't the final draft."

So there were other versions of this? Later versions?

"There were a few inaccuracies in that first draft," Goelz said.

Sauter interrupted with, "Let's go through and tick them off and make sure we're all straight."

"I'm not sure what I know, but certainly one that was indicated, that you raised, was the photo. And the truth on that was shortly before that meeting we did get the photos back. They had been given to somebody in Calverton. It was not a big deal. We just hadn't communicated it to the right person," Goelz concluded.

If it was such a minor incident, why did Goelz, the head of NTSB public affairs, not give a simple, direct response the day before? Probably because there were no real inaccuracies on which the NTSB could collectively hang its hat. But they needed something in order to cast a cloud of doubt on the entire ten-page memo.

Goelz was in front of the camera, still trying to

pitch the line that, "There were a few inaccuracies in that first draft." But he could only name one.

Sauter asked a question about the FBI and NTSB investigations not cooperating. Specifically, the NTSB report said the FBI turned over redacted interviews of witnesses. The FBI was blacking out portions of records being used inside the investigation before giving them to the NTSB, the lead investigating agency.

Goelz responded: "Even the slightest evidence of a criminal act would have pushed us to treat the investigation as a crime scene," he stated. The truth is, the FBI tried to take over the investigation in the early stages but was ordered not to by the Justice Department.

In true public affairs form, by the time Goelz finished answering the question, you would have thought there was an extraordinary level of cooperation between the world's two great investigative agencies. Somehow the redacted portions were forgotten, as was the almost open warfare in the press between the two agencies.

"I can assure you we have had complete cooperation from all federal agencies in this investigation," Goelz concluded.

What about the missile the FAA technician said was on a collision course with Flight 800?

"We saw those radar tapes shortly after we got on the scene. Our staff has reviewed those tapes, and they show absolutely no sign of a missile," said Goelz.

But the NTSB Chairman, James Hall, called for an explanation in writing for why the NTSB did not have access to the radar tapes in the hours after the crash. An FAA technician reviewed the tapes and preliminarily concluded there was a missile on a collision course with Flight 800. The FAA response acknowledged an "unexplained blip," and went on to editorialize that there was only a "remote possibility" that it was a missile.

There is evidence of a missile on the tapes. It's possible there is honest disagreement among experts about whether it is actually a missile seen headed toward Flight 800. But it is not true that the tapes "show absolutely no sign of a missile," as alleged by the NTSB head of public affairs.

By politically tagging it a "remote possibility," the door is not legally closed on it being part of a crime. That removes it from being accessed through the Freedom of Information Act for independent analysis outside the confines of the government.

We had confirmation not only of the document, but of the cover-up as well.

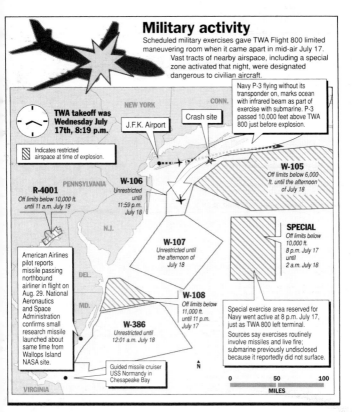

Military activity

Scheduled military exercises gave TWA Flight 800 limited maneuvering room when it came apart in mid-air July 17. Vast tracts of nearby airspace, including a special zone activated that night, were designated dangerous to civilian aircraft.

Navy P-3 flying without its transponder on, marks ocean with infrared beam as part of exercise with submarine. P-3 passed 10,000 feet above TWA 800 just before explosion.

TWA takeoff was Wednesday July 17th, 8:19 p.m.

Indicates restricted airspace at time of explosion.

NEW YORK

CONN.

J.F.K. Airport

Crash site

W-105
Off limits below 6,000 ft. until the afternoon of July 18

R-4001
Off limits below 10,000 ft. until 11 a.m. July 19

PENNSYLVANIA

W-106
Unrestricted until 11:59 p.m. July 18

N.J.

American Airlines pilot reports missile passing northbound airliner in flight on Aug. 29. National Aeronautics and Space Administration confirms small research missile launched about same time from Wallops Island NASA site.

DEL.

W-107
Unrestricted until the afternoon of July 18

SPECIAL
Off limits below 10,000 ft. 8 p.m. July 17 until 2 a.m. July 18

MD.

W-108
Off limits below 11,000 ft. until 11 p.m. July 17

W-386
Unrestricted until 12:01 a.m. July 18

Guided missile cruiser USS Normandy in Chesapeake Bay

Special exercise area reserved for Navy went active at 8 p.m. July 17, just as TWA 800 left terminal.

Sources say exercises routinely involve missiles and live fire; submarine previously undisclosed because it reportedly did not surface.

N

0   50   100
MILES

VIRGINIA

Overview of military zones in waters along initial flight path of TWA Flight 800. (*Courtesy of* The Press-Enterprise/*Paul D. Rodriguez*)

Early in the recovery process, divers from local police departments were escorted away from the search zones by the U.S. Coast Guard. They were replaced by federal teams. (*AP/Wide World/Mark Lennihan*)

On July 24, 1996, the USS *Oak Hill* arrived as the official Navy command ship to coordinate all recovery efforts. (*AP/Wide World/Charles Winthrow, USN pool*)

USS *Grasp* (ARS-51) serves as primary support for Navy diving and recovery efforts. (*AP/Wide World/Charles Winthrow, USN pool*)

Assistant FBI director James Kallstrom, *center*, is flanked by NTSB vice chairman Robert Francis, *left*, and Rear Admiral Edward Kristensen, *right*, during one of the dozens of press conferences held in Smithtown, NY. (*AP/Wide World/John Dunn*)

Peter Goelz, Director of Government and Public Affairs of the NTSB. (*AP/Wide World/Wally Sanana*)

Kallstrom confers with James Hall, Chairman of the NTSB, during a press conference concerning new evidence supporting the "missile theory." (*AP/Wide World/Adam Nadel*)

President Bill Clinton talks to reporters on July 25, 1996 at New York's John F. Kennedy International Airport as First Lady Hillary Rodham Clinton and White House Chief of Staff, Leon Panetta, look on. (*AP/Wide World/Ron Edmonds*)

NTSB charts showing the original plane location of debris recovered from the crash zone as of August 13, 1996. (*AP/Wide World/John Dunn*)

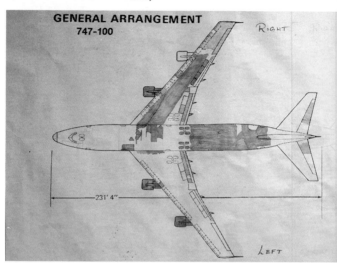

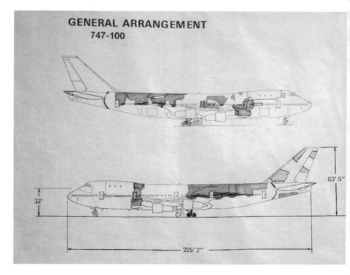

U.S. National Guard trucks carried wreckage to the hangar in Calverton, New York where it has been assembled. (*AP/Wide World/Mark Lennihan*)

Seats in the cockpit and just over the central fuel tank as arranged in the Calverton hangar. (*AP/Wide World*/CBS News/60 Minutes)

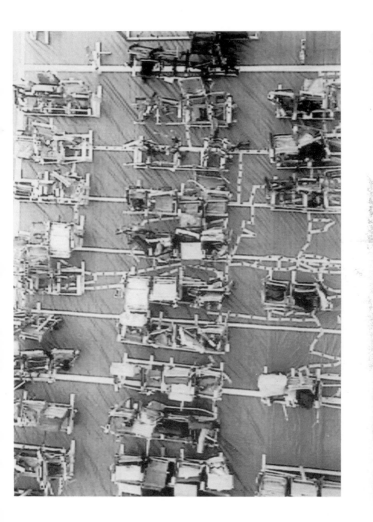

In this official NTSB photo, a significant portion of the middlebody of TWA Flight 800 has been reconstructed. (*AP/Wide World/NTSB*)

In this official FBI photo, debris is shown carefully laid out in the Calverton hangar as it is used to reconstruct the TWA 747. *(AP/Wide World/FBI)*

From the terrace of a seaside restaurant approximately fifteen miles from the crash zone, Linda Kabot took this photo between

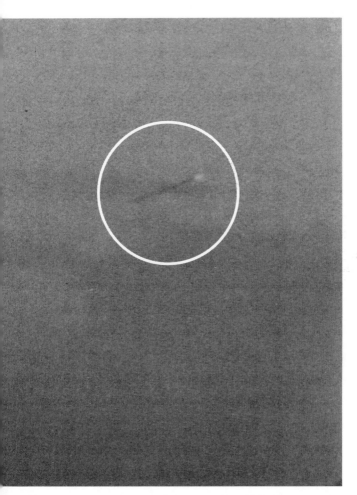

8:00 p.m. and 8:45 p.m. on July 17, 1996, which shows an airborne projectile or missile. (*Linda Kabot/SYGMA*)

The following debris schematic was compiled entirely from FBI and NTSB data. The pattern and distance of the debris clearly

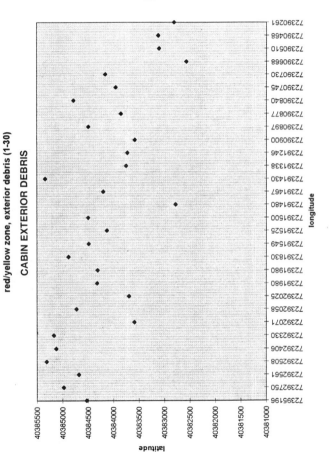

CABIN EXTERIOR DEBRIS

red/yellow zone, exterior debris (1-30)

indicate the impact of a nonexplosive missile, not an internal explosion.

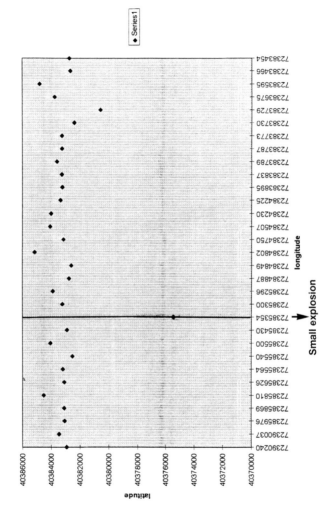

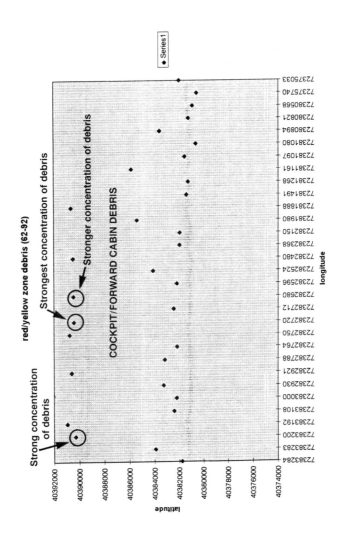

red/yellow zone debris (62-92)

Strongest concentration of debris

Stronger concentration of debris

COCKPIT/FORWARD CABIN DEBRIS

Strong concentration of debris

latitude

longitude

◆ Series1

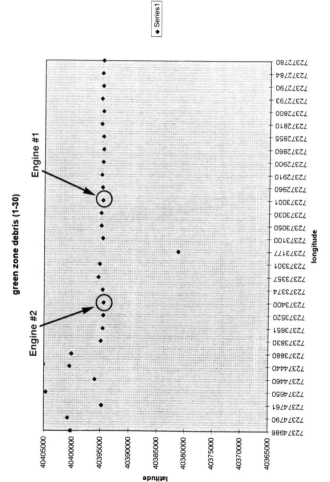

green zone debris (1-30)

Engine #1

Engine #2

latitude

longitude

◆ Series1

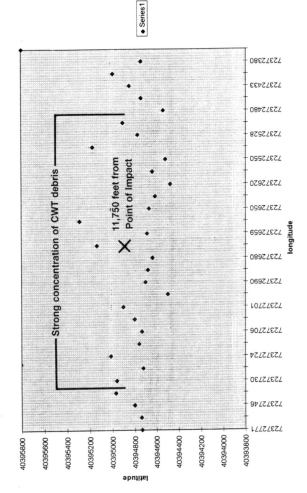

green zone debris (31-62)

Strong concentration of CWT debris

11,750 feet from Point of Impact

♦ Series1

latitude

longitude

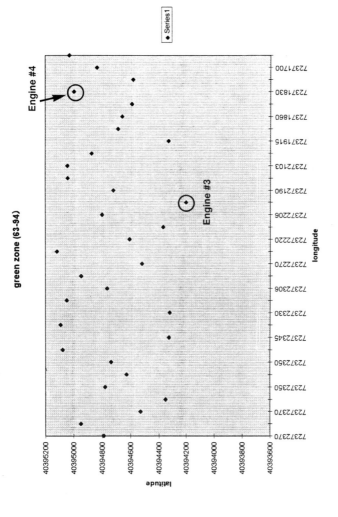

green zone (63-94)

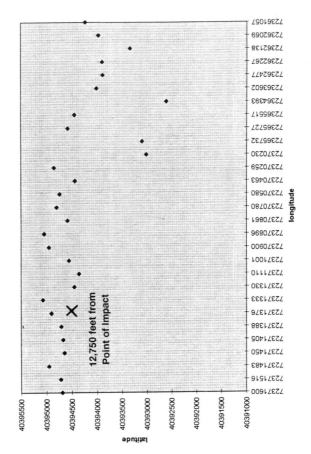

green zone (95-123)

♦ Series1

latitude

longitude

12,750 feet from Point of Impact

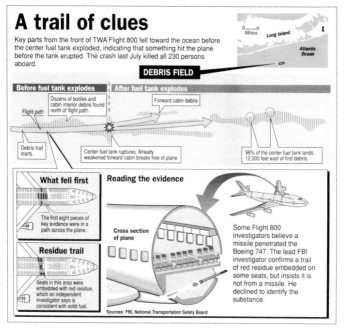

A trail of clues

Key parts from the front of TWA Flight 800 fell toward the ocean before the center fuel tank exploded, indicating that something hit the plane before the tank erupted. The crash last July killed all 230 persons aboard.

0—6 Miles   Long Island   Atlantic Ocean   N

DEBRIS FIELD

Before fuel tank explodes | **After fuel tank explodes**

Dozens of bodies and cabin interior debris found north of flight path.

Flight path

Forward cabin debris

Debris trail starts.

Center fuel tank ruptures. Already weakened forward cabin breaks free of plane.

98% of the center fuel tank lands 12,000 feet east of first debris.

What fell first

The first eight pieces of key evidence were in a path across the plane.

Residue trail

Seats in this area were embedded with red residue, which an independent investigator says is consistent with solid fuel.

Reading the evidence

Cross section of plane

Some Flight 800 investigators believe a missile penetrated the Boeing 747. The lead FBI investigator confirms a trail of red residue embedded on some seats, but insists it is not from a missile. He declined to identify the substance.

Sources: FBI, National Transportation Safety Board

Debris trail is one of the most powerful factors in indicating that TWA Flight 800 was shot down by a missile. (*Courtesy of The Press-Enterprise/Paul D. Rodriguez*)

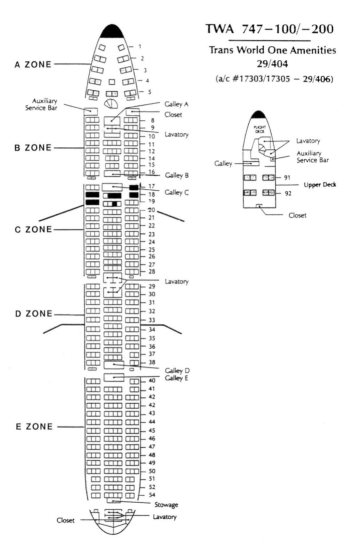

Schematic of the interior of TWA 747. Darkened area indicates the seats with reddish residue from missile exhaust.

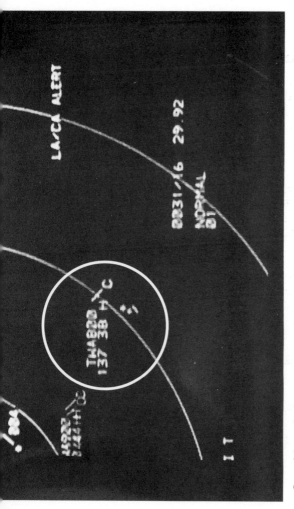

One of the radar images as recorded in an FAA air traffic control video tape. Circled items indicate TWA Flight 800 and an unidentified projectile in its path seconds before the explosion. (*AP/Wide World*)

Strip of seat padding showing reddish residue from missile exhaust.

CHAPTER TEN

The NTSB vs. The FBI

While Hendrix was trying to keep what he knew from being leaked before he had a chance to publish it, the Fire and Explosion team from the NTSB was doing just the opposite. Part of what they were trying to do was to make sure their theories got wide play to "support" the mechanical explanation the NTSB was trying to sell.

The Fire and Explosion team was headed by Dr. Merritt Birky, who had indicated that his group had yet to see any evidence of the erosion or pitting in any of the wreckage which would have been a telltale signature of an explosion. He also indicated that they had not discovered any static or fuel-transfer problems with the center wing tank.

Despite their lack of hard evidence, their first theory stated that a center wing tank fuel-pump must have caused the initiating event that led to the destruction of Flight 800. When the center wing tank pumps were brought up from the ocean floor, examined, and found not to have caused the crash, the theory changed the culprit to the center wing tank scavenge pump. But this was just a guess, elevated to theory, then to fact by leaking it to the media.

Since the initiating event in the destruction of Flight 800 occurred about fifty feet forward of these pumps, NTSB theory eventually worked its way forward to the front of the center wing tank: Static electricity in a fuel line that ran through the forward section of the tank became the new best guess. This was virtually impossible to disprove. It was the perfect mechanical theory when a cover-up is in progress. You can't prove it happened but, more importantly, you can't disprove it either, unless of course you conclusively prove something else caused the crash.

The Red Zone debris field had been sanitized in the opening hours of the investigation. Missile parts and fuselage debris with red residue disappeared from sight on trucks. Once the working investigative shop was set up inside the hangar, FBI agents routinely removed parts from the floor of the Calverton hangar without logging them out. Thus, if there were other indicators on the debris of an explosion or of some other cause, the FBI's unrestricted removal of the debris in question from the custody of independent NTSB investigators compromised the whole process. In this way, any evidence related to theories not consistent with government public affairs statements would conveniently disappear, and those theories deemed to no basis in fact.

However, there is absolutely no evidence that static electricity was involved in the plane's demise. In fact, the opposite is true. Birky and his team searched for evidence of static electricity and "have not discovered any static or fuel-transfer problems with the center wing tank," according to the NTSB chairman's November 15, 1996, report.

The chairman didn't say there was a little proof and they were on the trail of more. He made an absolute statement: There was no proof whatsoever. But this didn't stop someone at the top of the NTSB from leaking this already debunked theory to the media. It would be reasonable to wonder why news organizations dealing with the NTSB since July on this story, especially NBC, now on its fourth iteration of unsubstantiated mechanical theories, continued to pump out stories with no independent foundation to back up the high-level "leaks."

Hall's November 15 report also revealed that "No evidence was found of fatigue in fuselage structure for pieces associated with the Red Zone." So, four months after the crash of Flight 800, there was absolutely no indication of a mechanical problem. Similarly, the bomb theory is greatly underplayed. It is not even mentioned in the chairman's ten-page report. But a missile is mentioned, twice. Ron

Schleede indicated that geosynchronous satellite sensors may have picked up either radio transmissions or infrared emissions related to this accident.

The report further states that Ron Schleede would write a letter for Bernie Loeb's signature to Ron Morgan for a full explanation of the FAA handling of Air Traffic Control and radar tapes concerning TWA Flight 800. The letter would reference the technician who did the analysis resulting in conflicting radar tracks that indicated a missile. It would also inquire as to why that information was reported to the White House and sent to the FAA Technical Center before the Safety Board was given access to the data.

The FAA eventually responded to this letter, saying that further analysis did confirm an "unexplained blip," but that there was only a "remote possibility" that it was a missile. The letter does not say why the possibility was "remote."

The reference to possible "infrared emissions" is particularly interesting in light of the revelation that the Navy P-3 aircraft passing over the scene of the shootdown at 20,000 feet, just prior to Flight 800 being hit, was carrying infrared equipment to monitor a military exercise in warning zone W-105. Infrared detects heat. The P-3 was overhead to monitor military weapons that create heat.

A high-ranking military source has confirmed that there was a drone in the area. Another source says that a U.S. Army reserve unit was out toward the eastern end of Long Island the night of the shootdown. It had been involved in military exercises earlier in the week that included the use of drones. All of which implies strongly that the infrared-carrying P-3 was there to monitor a military exercise that included a drone.

Working on the assumption that the military did not just stand around watching the drone flying aimlessly about warning zone W-105, it is reasonable to assume they intended to shoot it down. They'd spent billions of dollars on high-tech radar, computers, and missiles to perform that act.

Dr. Hall was also concerned about the FBI's lack of cooperation with the NTSB side of the investigation. Sources inside the Calverton hangar said the NTSB first learned about possible "bomb" residue inside the 747 when they heard it on the television news. The sources describe two virtually separate investigations. Hall's ten-page memo confirms this.

Due to the initial information that this investigation would at any moment turn into a criminal investigation, NTSB investigator Norm Weimeyer had access only to witness statements and was not able to take notes or prepare summaries of the interviews. Redacted statements prepared by FBI agents for ground personnel involved with the dispatch of Flight 800 have recently been provided to the NTSB; however, a review indicated that the proper accident investigation questions had not been asked. A review of these statements is currently underway, and a schedule is being set up to reinterview many of these people. The chairman directed that if there were any resistance from the FBI, the investigators were to get it in writing. These are certainly not the words of someone who believes the NTSB and FBI are "fully cooperating" in a search for the truth.

The chairman of the NTSB is responsible for an investigation that allegedly cannot find one single piece of relevant evidence, even though about ninety-five percent of the wreckage had been recovered by November 15, 1996. Theories unsupported by evidence poured from the NTSB side of the investigation.

The November 15 chairman's report is actually an exercise in damage control:

> Hearings may take place concerning the investigation as soon as February 1997 and the board should be in a position in which as little criticism as possible can be levied against it.
>
> We should anticipate the need to defend every decision we make.

Less than two weeks after the chaiman's resport was written, there was a meeting of all NTSB teams at the Calverton hangar. New personnel from the NTSB's Washington, D.C., office arrived and immediately brought Dr. Birky and his team under attack, according to a source who attended the meeting.

All theories were put back on the table because Birky had absolutely nothing to back up any of the mechanical failure theory except the fact that, at some point in the chain of events that would destroy Flight 800, the center wing tank exploded. Of course the right wing tank also exploded, but that knowledge did not lead to a succession of unsubstantiated theories about how this triggered the loss of the plane.

This new NTSB objectivity lasted only a few days before the leaders of the insurrection were put in their place. Then as expected, the *NBC Evening News* presented a lengthy analysis of what appeared to be the cover-up's final stand: Static electricity.

CHAPTER ELEVEN

HARD EVIDENCE: MISSILE RESIDUE

The single most important development in my investigation came early in the new year. Staring down at the envelope, I knew I had the proof I'd been looking for for months. I ripped off the top of the package and saw two small swatches of foam, both of which were covered with a reddish-orange residue. Hangar Man had, on his own, come through with material from the seats, along what I knew to be the missile path inside the forward cabin.

But merely having swatches of foam in one's custody meant nothing. It was what was on that foam that was important, and how quickly I could get it analyzed, broken down into its component substances, and the results of the lab test disseminated to Dave Hendrix at the *Press-Enterprise*, to Mark Sauter at *Inside Edition,* and to any others who wanted to know about what had really happened.

Now that I had the residue, I wasn't about to FedEx it myself. I would hand it over to a lab facility. But that didn't mean that I had to wait, not with major news organizations struggling to find out what was really going on inside the Calverton hangar.

So I hand-delivered the swatches to a technician at West Coast Analytical Services, Inc., a California chemical testing laboratory, and told them all I wanted was an analysis that yielded a list of the chemical components of the residue. I wasn't looking for what it was, at least not from the testing lab. I merely wanted to know what chemicals the residue contained, and in what quantities, and I would take it from there. My idea was to take the analytical report over to a number of rocket fuel manufacturers like Hughes and Thiokol to find out if the material in the analysis was consistent with the residue from a solid fuel rocket.

Antiaircraft missile solid fuel propellant generally contains:

1. Metal fuel, usually aluminum powder.
2. Oxidant, usually a perchlorate such as ammonium or nitronium perchlorate.
3. A synthetic rubber binder.

So aluminum is the first signature of a solid fuel. Perchlorates are generally only used in pyrotechnics and rocket fuel, making it an ideal solid fuel signature to look for. Calcium is used in conjunction with perchlorate in solid fuel, calcium producing the flame and perchlorate the oxygen. Unfortunately, perchlorates are water-soluble. It is impossible to get a reading on perchlorates from solid fuel propellant residue that has been immersed in the Atlantic Ocean for at least two weeks prior to testing. So the relevant signature here would be the calcium.

Silicon and copper are the third signature of solid fuel. Silicon is the key ingredient in synthetic rubber, when combined with a small amount of copper. The synthetic rubber, while in a liquid form, receives the other ingredients: perchlorate, calcium, and aluminum powder. They are mixed into synthetic liquid rubber, then poured into a container where they rapidly harden before the ingredients can settle to the bottom. Trace elements of metals used in the construction of a missile are also possible signatures to be tested for—hardened steels and metals such as titanium.

A sample that has been immersed in salt water will test positive for magnesium, making it impossible to determine whether the magnesium comes from a source other than ocean salt.

West Coast Analytical Services, Inc., received a sample of the reddish-orange residue on January 23, 1997. The initial test was for "traces of aluminum using emission spectroscopy." Although aluminum was the focus, all metals and other compounds would be detected and reported. A second sample was held in reserve, pending the outcome of this test.

The results were:

Magnesium	18	%
Silicon	15	
Calcium	12	
Zinc	3.6	
Iron	3.1	
Aluminum	2.8	
Lead	2.4	
Titanium	1.7	
Antimony	0.53	
Nickel	0.38	
Manganese	0.21	
Boron	0.081	
Copper	0.053	
Silver	0.032	
Chromium	0.032	

All of these elements are consistent with residue from a solid fuel missile.

Numerous unnamed government sources have told the press that these elements are consistent with glue. Other unnamed sources from within the federal government have leaked the claim that these elements are consistent with other substances. They have never said what these substances are, however. Outside the government's faceless, nameless debunkers who leak stories without any corroborative evidence, no one has been found who will publicly state that these elements are consistent with any other substance that could be found on airline seats.

Three journalists are known to be researching the issue of glue used in the manufacture of airplane seats. David E. Hendrix with *The Riverside Press-Enterprise,* Bill Scott, *Aviation Week,* and Reed Irvine, Accuracy in Media, Inc. Their research to date is essentially the same: Little or no glue is used in the manufacturing process of airline seats. Aside from a small amount used to join two or more pieces of foam together to make the headrest, Velcro and other fasteners are used to hold the various parts of the seat to the frame. Because of the heavy usage and need to frequently repair or replace portions of the seat, the fabric and other nonmetal parts are designed to be broken down in modular units to avoid the time and expense of replacing an entire seat.

Missiles, on the other hand, do use adhesives, or glues. According to Thiokol, one of the oldest builders of missiles in the United States, "Adhesives play an important role in joining the various dissimilar materials of which the nozzle is made, in assuring the soundness of joints, and in sealing spaces between materials to prevent gas leakage. Sealants such as zinc chromate and silicone grease may be used to plug cracks."

CHAPTER TWELVE

OUT OF CONTROL

National Airport: Level B-4 in long-term parking overlooks the airport Metro stop and the covered walkway leading to the parking area—A great place to meet government informants. I watched from the top level as my informant, Hangar Man, slowly made his way on foot to the parking garage. It was January 1997, the middle of a cold, East Coast winter.

Our conversation began quickly. He told me that the FBI missile team "felt strongly it was a missile." The two-man team leaned toward a "third-generation" infrared (IR), shoulder-launched anti-aircraft missile, which has the range to reach a 747 at 13,700 feet according to government missile experts. Earlier IR shoulder-launched missiles commonly available on the black market are not up to the job, at least not at an altitude of 13,700 feet. But a third-generation missile leaves the door open for something other than friendly fire. A third-generation missile would not lock on the exhaust heat behind the aircraft, the informant said. Instead, it would search for a heat signature and then focus on "the center of the heat." In the case of a 747, that mass is forward of the centerline of the wing, according to the informant.

The FBI and NTSB investigations are highly compartmentalized and it is unlikely that the FBI missile team would have direct knowledge of missile parts and fuselage debris with reddish-orange residue being trucked away, early in the process. Nor would the FBI missile team have access to the radar tapes to determine for themselves whether a missile was on a collision course with Flight 800. What they do know is that it would be fatal to their careers if they went against policy and pushed the belief that a missile with a semiactive radar brought down Flight 800. It is a very short step from there to "friendly fire."

No federal employee could do such a thing without first making the very difficult personal decision to take on the might of the employer who feeds their family and become a whistle-blower. Congress has attempted to pass laws making it possible for people within the bureaucracy to reveal corruption and cover-up. The sad truth, however, is that the bureaucracy is vastly more powerful than congress. In spite of their best efforts, congress has not been able to stop the government from persecuting and destroying those who dare to take them on.

There is a second type of missile the FBI team talks about when discussing their work with other investigators at the Calverton hangar: semiactive, radar-guided. A member of the FBI missile team told my source that a missile with an internal radar-tracking device could be "seen" by FAA radar.

Government missile experts told the FBI that FAA radar would pick up the radar signal. But the same radar could not see an IR missile. Since the FAA technician in New York who analyzed the radar tape shortly after Flight 800 crashed saw a missile on a collision course with the 747, there is a strong possibility it had a radar guidance system. Such a missile almost certainly mandates a government delivery system, which in turn strongly suggests friendly fire. And the FAA technician's analysis was confirmed by the FAA Technical Center, which said there was an "unexplained blip." When this was revealed in a FAA letter, the obligatory commentary was inserted that there was only a remote chance it was a missile. Since this commentary is not backed by any evidence of analysis, and flies in the face of two separate analyses, the commentary carries little weight.

Since friendly fire is a subject that cannot, under any circumstances, be considered by a government investigator, NTSB or FBI, a non-IR missile is the only projectable option discussed.

According to the source, the investigation itself had

reached the envelope of what could and could not be officially explored. For example, more than six months after Flight 800 crashed, the NTSB had not decided what firm would attempt to analyze the sound heard on the black box just before it lost electrical power.

"Why?" I wanted to know.

"No one knows," was the response.

It should have been one of the NTSB's top priorities back in July 1996. The NTSB has checked it against sounds heard on the cockpit voice recorders of other doomed planes, such as the Pan Am 747 brought down over Scotland a few years earlier. But the Flight 800 sound is different from anything on record.

It is faster than the others. A bomb will send a vibration through the plane's structure at about 350 feet per second. The sound recorded on Flight 800's cockpit voice recorder was traveling in excess of 2000 feet per second. This fact, and the lack of any bomb explosives found in the wreckage, should have completely eliminated the bomb theory from consideration. Also, it is very doubtful that any "mechanical" could send shock waves through the fuselage at a higher rate of speed than a bomb, because there would be no explosive or impact energy involved. However, the sound of a missile with an inert warhead impacting a plane is not on file. And the NTSB is in no hurry to add to their body of knowledge in this area.

I asked about the Suffolk County coroner. He had

been a long-term problem for the NTSB because of his insistence that damage to victims bodies did not correlate to any NTSB mechanical theory. Hangar Man's most recent information was that the coroner's position had not changed. The center wing tank could not have been the initiating event because of the lack of lower leg damage.

My informant also said that Jim Wildley, head of the NTSB Sequencing Team, had too many anomalies that could not be explained. The Sequencing Team had attempted to develop a logical sequence of events in the breakup of Flight 800 that would correspond to the NTSB mechanical theory.

What I was hearing was the description of a government document outlining how the evidence could be altered in order to prove a mechanical cause. So I asked about the availability of the NTSB Sequencing Team documents outlining the latest theory. An ongoing problem from my side of the investigation was the myth that NTSB theory is in any manner based on sound reasoning and judgment. The NTSB's own written words outlining a conspiracy to thwart justice was the best possible response to the never-ending line of public affairs spokespeople.

The informant thought the documents could be obtained. Going down my checklist, I stopped at FBI residue analysis. The informant said an NTSB team had been pushing for months to obtain the FBI lab test results from a sample of the reddish-orange residue found exclusively on the back of seats in rows

17, 18, and 19. The FBI took samples from the backs of the following seats:

Row 17, seat 8

Row 18, seats 6 and 7

Row 19, seat 7

The FBI was stonewalling, my source told me. They weren't saying there was a delay. They just weren't responding to a very simple, reasonable request for information the FBI had promised to share with the NTSB side of the investigation. Nor was the information being shared with the FBI teams at Calverton.

When individual investigators at Calverton continued to press for the analysis, the FBI didn't say that it was glue or something else seen many times in the past, and common to airplane crashes. Nor did they say that the residue was consistent with something other than solid fuel from a missile. They stonewalled. At the hangar, nonresponse was interpreted as bad news, indicating a probable missile signature. There is no other reasonable explanation, the informant said. A few weeks later, the FBI finally admitted why they would not give the NTSB the results of their analysis: They saw it as evidence in their criminal investigation.

The informant also said an NTSB team recently reviewed the statements of New York Air National Guard 106th Rescue Wing pilots Major Fred Meyer and Captain Christian Baur, who were piloting an HH-60G helicopter over Long Island the evening of July 17, 1996. According to a 106th Rescue Wing press release, the two helicopter pilots who saw TWA Flight 800 explode on the night of July 17 have not changed their statements. Both witnessed a streak of light moving from east to west prior to the initial explosion. Both pilots, who have made statements to the FBI, do not know what caused the streak of light.

According to the informant, the press release only hinted at the truth. Captain Baur was ordered not to say *missile* during the interview, the informant suspected, because of the manner in which the statement was given. Nor could he say the burning object hit Flight 800. Instead, Baur said a "burning object," going from east to west, hit another object and caused it to explode. So two Air National Guard pilots can be added to the three dozen highly credible witnesses who saw something travel through the air and hit TWA Flight 800, causing it to crash.

By March 1997 the FBI authorized Major Meyer to be interviewed by a journalist from Aviation Week, *but only if monitored by an Air Force public affairs officer. During the interview, it was publicly revealed for the first time that Captain Chris Baur had witnessed something, traveling from east to west, impact Flight 800. For all these months the 106th had been putting out false information in its press releases, which alleged that the two pilots, Major Meyer and Captain Baur, had seen only a streak of light. It was suggested that it may have been a meteor, when in fact the military knew that Baur had seen something traveling from east to west. He saw it slam into Flight 800, and he saw*

Flight 800 break up in flight and crash into the ocean. The two officers were so close to the crash that they saw several bodies fall into the ocean.

It would be unreasonable to believe or even suspect that the 106th New York Air National Guard would, on their own, withhold such information. They must have been ordered to. Sources inside the investigation have long said that the FBI asked the 106th to prohibit Baur from speaking out. That request became an order when the FBI leaned on the New York Air National Guard command.

Meanwhile, another story was breaking about the mysterious blip that appeared on the FAA tapes at the time of the Flight 800 disaster.

In December, the *Washington Times* reported that a missile expert from the Defense Intelligence Agency (DIA) had stated to selected staff members in Congress that he believed a shoulder-fired missile brought down TWA Flight 800. "The same DIA official told staff members that he personally was called in by the FBI in the days following the explosion to assist with witness interviews."

At the time, the two-man FBI team investigating the missile theory supported the observations of the DIA expert.

It caused quite a flurry of activity in the investigation for a short period of time. According to *Newsday,* "Investigators probing the crash have fired missiles at targets on top secret military bases in California over the past few weeks. Missiles were reported to be fired at aluminum sheets, but caused more damage than any found on the Flight 800 wreckage."

By all reports, the new direction made this investigation the first in history of a commercial plane shot down by a missile. According to one *Newsday* source, unarmed, small-scale missiles caused the same kind of damage found in the hangar at Calverton.

The official public affairs statement concerning this line of testing included a complete denial of the involvement of U.S. missiles. The missile tests concentrated on the possibility of a terrorist missile.

CHAPTER THIRTEEN

THE STORY FINALLY BREAKS

By March the press had been neutralized by the steady pounding of the word "mechanical," put out by the small army of public affairs officers and unnamed sources swarming all over the press and handing out the same line over and over. In-depth articles seldom appeared, only press releases, and nothing ever turned up in the papers with substantive facts behind it. So it must have been quite a surprise when, on Friday morning, March 7, 1997, FBI public affairs received a call from David Hendrix, a reporter for *The Riverside Press-Enterprise* in Southern California, about a story his paper wanted to run about the strange residue in Flight 800's cabin.

1530 hours, at *The Riverside Press-Enterprise:* It's more than eight hours after David Hendrix told the FBI

he is running a story saying there is an apparentl
red residue trail inside the fuselage of Flight 800 tha
tests out as missile propellant. Finally the callbac
comes. Jim Kallstrom, assistant director in charge o
the TWA Flight 800 investigation, wants to talk t
Hendrix.

Sitting at his computer terminal, possible interviev
questions on the screen, Hendrix begins to talk witl
Kallstrom.

HENDRIX: I'm doing a story saying there is ar
apparent residue trail through TW/
Flight 800 that tests out as missile pro
pellant. I have independent confirma
tion that you had residue in you
possession no later than August 3 anc
that the trail across the fuselage wa
completed at Calverton no later thar
the end of August.

*About halfway through the opening statement it wa
apparent that Kallstrom had turned on his speaker so other
in the room could listen to both sides of the conversation.*

KALLSTROM: It's not true. We've looked over ever
piece of this plane, left and right, witl
the best experts in the world. If I hac
proof, at that moment I would hav
stood up and said it to the world. I'n
confident that it is not true.

There is a red residue trail. It has no logical connection to a missile. I'm not going to get into it. There's a logical explanation, but I'm not going to get into it.

The notion that you would write an article saying this is proof of a missile . . . there's no basis in fact. To my knowledge that is not factual.

HENDRIX: The FBI took tests from seats in rows 17, 18, and 19. What were the results of those tests?

KALLSTROM: We're not in the habit of discussing lab tests.

Hendrix began to read the list of elements from the West Coast Analytical Services, laboratory report.

KALLSTROM: Are you reading from a document?

HENDRIX: Yes.

He declined to comment any further on this topic. The interview continued.

HENDRIX: A story will soon be breaking from Europe about FAA and satellite imagery showing what appears to be a missile headed toward Flight 800. Three radar sweeps show a missile at 1500 to 2000

	miles an hour intersecting TWA 800 Does the FBI have such documentation?
KALLSTROM:	I've heard it talked about on the Internet. I've heard all that chatter Nothing to it. We've analyzed them til the cows came home. [We] have analyzed every possible sensor.
HENDRIX:	Does the FBI still consider the 17 July FAA radar images showing a possible missile on a collision [course] with TWA 800 an anomaly?
KALLSTROM:	I've never talked about an anomaly.
HENDRIX:	Has there been an investigation of why FAA tapes were routed to the White House and FAA before going to the NTSB for analysis?
KALLSTROM:	That's absolute nonsense. To my knowledge [there were] no circuitous routes.

The NTSB's own documents reveal this routing of the FAA radar tape information.

HENDRIX:	The FBI has thirty-four highly credible witnesses. Our sources say you took each one to a site and triangulated from where the shot would have come from and it was out of warning area W-105 and intersected Flight 800. Comment?
KALLSTROM:	I don't know.
HENDRIX:	Have the analysis and computer

156

KALLSTROM:
HENDRIX:

enhancement of the Kabot photo, and a second photo showing a white trail going into the air, been completed? What have they shown?

KALLSTROM: I don't know if enhancements [have been] done or completed. Can't answer what they may have shown.

HENDRIX: The FAA's own records show Navy exercises were scheduled and sources say they were actually conducted the night of 17 July. Has the FBI independently determined who and what was involved in those exercises?

KALLSTROM: [I'm] not at liberty to discuss any of that.

HENDRIX: Was a drone in the area when Flight 800 went down?

KALLSTROM: We looked early on about the possibility of a drone. [We have] no information that there was a drone in the area that was used or shot at by the military.

NEW DATA SHOW MISSILE MAY HAVE NAILED TWA 800 stretched across the top of the March 10, 1997, edition of *The Riverside Press-Enterprise*. The left-hand side of page one outlined the case for a missile shootdown and the right-hand side carried the government rebuttal.

Loren Fleckenstein, a young journalist with the *Press-Enterprise*, opened the debate with an article entitled, "Debris Pattern Provides Key to Mystery: The Answer to the Mystery of What Happened to Flight 800."

Hendrix opened the right-hand article, entitled, "But FBI Rejects Key Portion of Conclusion," with an overview of the issues including the missile residue, then went into the government's position:

James Kallstrom, the FBI's assistant director in charge of the investigation, confirmed the existence of the residue but declined to identify the substance and said it did not point to a missile.

"There is a red residue trail but it has no connection to a missile," Kallstrom said Friday in an interview that he said was breaking his own rule against discussing specific evidence. "I'm not going to get into it. There's a logical explanation but I'm not going to get into it."

NTSB spokesman Peter Goelz said there was no physical evidence supporting a missile or its propellant.

"We'll be testifying before congress on Tuesday that as of today there is no physical evidence of a bomb or a missile in any of the records (evidence) that we have recovered," Goelz said Sunday.

Although FBI and NTSB investigators have repeatedly said that tests do not show any hint of explosives from a bomb or warhead, they had not discussed propellant or solid fuel, used to power even unarmed practice missiles.

A source inside the investigation, who spoke on the condition his name not be used, said reconstruction and crash experts working at a former Grumman Aviation hangar in Calverton, NY, have waited for the FBI to report on lab tests on the red substance embedded in the seats.

"That's the frustrating thing," the source said. "These samples were taken in August and no FBI results have been shared at this level."

He said another investigator brought it [the trail of red residue] to his attention. Requests in hangar meetings for an analysis have been thwarted by FBI statements that the material is part of the criminal investigation and can't be discussed.

Hendrix's article went on to describe two experts who analyzed different parts of the evidence for the *Press-Enterprise:* " 'Things like magnesium and aluminum make the missile work,' said a Hughes Missiles engineer, who agreed to speak on the condition that his name not be used. 'It depends on how fast you want the motor to burn. There's lots of aluminum. Aluminum burns faster. It raises the temperature and gives more thrust.' "

Hendrix went to "A longtime FAA investigator and attorney not connected to the crash [who] said that his analysis of the NTSB crash documents made him believe some piercing object entered the plane and traveled right to left, seriously damaging the plane and generating fierce winds inside, created by the partial vacuum caused by 400-mph winds passing by the left-side hole. The dynamic forces would have sucked people and objects out the expanding holes that soon merged from top to bottom, the investigator-attorney said."

Hendrix also interviewed Richard Russell, a retired airline pilot who has also had extensive experience as a crash site investigator with an airline pilots union. Russell "said he has copies of FAA radar tapes showing a smaller object on a collision course with Flight 800 seconds before the plane began disintegrating." The object appeared to be traveling at 1500 to 2000 miles per hour, Russell said.

The Hendrix article announced Russell's intention to make the radar tapes public within a week after the *Press-Enterprise* article went to press. That was on Monday.

On Tuesday, however, "The FBI seized a Federal Aviation Administration radar tape that purportedly showed an object speeding toward TWA Flight 800 seconds before the plane exploded," the Associated Press reported.

The FBI had known for months that Richard Russell had the FAA tapes. He didn't keep it a secret. The Government put out its usual statements questioning their authenticity, and the press continued to snooze. But now there were two threats to be faced simultaneously: the FAA radar tapes and the missile exhaust residue. Having both at the disposal of the press might create a feeding frenzy that could not be controlled.

Tuesday, the day after the *Press-Enterprise* story broke, I was completing my preparations to find a much quieter spot to finish this manuscript. My published phone number never stopped ringing. It was impossible to write. When I heard about the confiscated Russell tapes, I knew I was next. After all, only I knew where the remaining missile residue sample was. The FBI was fanning out across the country looking to pull back whatever evidence it could. They would soon be coming over to my house to look for any signs of where I might be. So I caught the few hours of sleep I could, packed off the dog to a kennel, overnighted the residue sample to a producer at CBS, who agreed to take custody for testing purposes, and left for a quiet location where I could finish this book.

I had been right. The next day, Thursday, FBI and Department of Defense investigators were all over Saint George's Hundred, the quiet, somewhat rural neighborhood where my family had lived for twenty years. FBI agents spotted a Federal Express package sitting on the covered porch. They read the bill label. It was from Christina Borjesson, a producer with CBS News. I had loaned her some documents and they were being returned. It was a lead. The FedEx label gave the FBI the clue they needed. Someone called New York on a cell phone and the agents were off.

One day later, my wife, Liz, flew into Norfolk airport from her office at the St. Louis TWA hub. She was going home to get the dog and relocate our son, Eric, due to return home from college for spring break. We didn't want him to walk into an empty house guarded by federal agents.

Liz's flight landed at Norfolk and as she walked off the plane, a person who works at Norfolk airport pulled her off to the side, away from the other passengers, and asked her if she was Liz Sanders. When she said yes, this person quickly hustled her down the back stairs of the terminal to a quiet location, out of sight of everyone, and told her that the FBI was waiting for her. The FBI was over in the terminal, this person said, and were complaining that they couldn't find her husband, Jim, and wanted to know if he had recently flown out of Norfolk Airport. Then they wanted to know when Liz's plane was due to land.

Within seconds, airport personnel were alerted to what was going on and ran interference. It was as if the resistance had quietly begun. The FBI didn't get to the plane in time to see her exit stage left. Friends and coworkers whisked her out of sight, put her in a limousine, and had her driven to Williamsburg, to a hotel where some of our son's friends worked. The owner, who had never met her, immediately provided her with a room. Eric's friends pitched in and did what Liz was not able to do—they contacted Eric and got the dog.

By this time, Liz was in contact with a New York lawyer we'd retained in an attempt to afford us some legal protection. The lawyer was a former federal prosecutor and knew the FBI agent in charge of the Sanders case, Ben Campbell. Campbell admitted that there was a subpoena out for me, and that they wanted to question Liz.

Campbell was advised that I was protected by the federal Shield Law, which gives a journalist modest cover when investigating federal crime and corruption. But family members of a journalist are afforded no such protection.

The pressure would continue.

CHAPTER FOURTEEN

Conduct Unbecoming

As far back as October 1996, just a couple of months after the crash of Flight 800, the inherent weakness in press investigations had become painfully apparent to me. It wasn't that the investigations themselves were flawed, but that there had been what looked like a paradigm shift in the way journalistic coverage was conducted over the past few years, so that it was tougher to get a story out in the traditional mainstream press than it had used to be.

There seemed to be a climate of fear in the air, fear of government reprisals, fear of lost ratings because the news was unpopular, fear of exposing "hard news" to too many people.

Newspapers themselves were fighting to stay alive, faced with competition from locals and low-overhead regionals, tabloids, and even national news magazines which presented digests of the news in easily swallowable nuggets. None of this is meant as criticism, only a description of the kind of landscape through which I felt I had to walk the more I realized I was on the trail of a story that nobody wanted to break.

Without becoming too disingenuous in criticizing it, I think journalists have become so fixated upon instant news and the constant new lead story, they've become reluctant to rely on the building-up of a base of facts, continually analyzing and drawing from that base, adding to it when evidence is confirmed, and withdrawing it when it is incorrect.

What's unfortunate is that the press is letting federal investigators avoid serious public scrutiny by not aggressively demanding an accounting of what's gone on in the hangar for the past eight months. And the only real truth to emerge from the hangar is that the press has been used and abused by the official spokesperson here on an almost daily basis.

However, these observations are not limited to the print media. The major networks seem to have become the government's mouthpiece when it comes to delivering the confused and conflicting accounts of the downing of TWA Flight 800. In fact, some TV news organizations, when confronted with the opportunity to report significant and potentially controversial items related to this case, have been reluctant to take a stand.

After I received the swatches of seat foam from my

source at the Calverton hangar, one of the first people I told was Mark Sauter, my friend and fellow journalist at *Inside Edition,* who has been instrumental in confirming the NTSB report. His initial response, as I expected from a reporter with his aggressive style, strongly suggested that he was interested in the story. The next day, however, Mark called to convey the official reaction he'd received from *Inside Edition.* Sounding as if he were reading from a legal document, Sauter conveyed a warning to me not to engage in any activity connected to residue currently under the control of the federal government that had been removed from government control, at least not while I was in any manner engaged in activity on behalf of *Inside Edition.*

Shortly afterward, I alerted *Inside Edition* by fax that I had gained access to the residue samples from a protected source. The next day I received the following brief letter:

At this point, we have decided not to continue your project. Therefore, you are no longer working as a freelance producer for *Inside Edition.*

It was a catch-22 situation. The only hard evidence to prove the missile theory was the missile residue. But despite the time-honored journalistic tradition of gaining access to sensitive information from a protected source, the fact that I had this proof sent the very people whom I needed to break the story scurrying for cover.

Several months later, it happened again, only this time it involved CBS News and cost me the control of the last swatch I had. I had been working with Christina Borjesson, a producer at CBS News in connection with a possible report by *60 Minutes* about the missile theory. I had sent the only remaining piece of seat material to CBS with the understanding that it would be independently analyzed and the missile theory settled once and for all.

However, that's not what happened.

Apparently, the agents that had visited my Virginia home got Borjesson's name from a FedEx envelope left at my front door and contacted her. On Tuesday, March 18, 1997, representatives of the NTSB requested the piece of seat material from CBS News. In a remarkably quick exchange, with little resistance at all from the news organization, and at the insistence of the network's attorneys, one of the senior producers handed over the material to the federal authorities. The last bit of concrete evidence outside of government control supporting the missile theory had effectively vanished. As a bona fide journalist and source whose life had been turned upside down while investigating a story of national importance, I was not even consulted in this decision.

When this breach of journalistic privilege was reported in the *New York Post* the following day, CBS at first denied it had ever happened. Then, there was a series of public statements designed to distance the news organization and the network from the controversy. Finally, they issued their formal position: The material was given over to the federal authorities because CBS did not want to be perceived as impeding a federal investigation as important as the TWA Flight 800 disaster.

From corporate news organization that had betrayed its relationship with a shielded journalist and source to public servant interested only in helping its government solve a painful case in which 230 people lost their lives—CBS managed to put a positive spin on actions that may have prevented the American people from ever finding out the truth.

Despite the actions taken by both major news organizations, the events indicate a pattern that strongly suggests a concerted effort by the federal authorities to cover up all independent avenues in support of the missile theory.

5 USC, section 552, subsection (b)(7)(A)(7) of the Freedom of Information Act covers security classifications of records or information compiled for law enforcement purposes, but only to the extent that the production of such law enforcement records or information could reasonably be expected to interfere with law enforcement proceedings.

The FBI and the NTSB have confiscated information and evidence twice during the course of so many weeks which allegedly could obstruct the cause of justice and the investigation of the downing of TWA

Flight 800. The first time, the FBI managed to get a grand jury subpoena to take possession of the Russell copies of FAA radar tapes. When the existence of these tapes became known in the early stages of the investigation, the FBI stated publicly time and again that these tapes contained no conclusive evidence supporting the missile theory, or for that matter, any of the other theories relating to the disaster. The tapes were of no value to the investigation.

If this was the case, why confiscate them?

Once the existence of the reddish-orange residue on the seats of Flight 800 became public knowledge, the FBI and the NTSB went great lengths to discredit this evidence. Their analysis first stated that it contained elements consistent with "glue." Then, unnamed government experts stated that it contained elements consistent with a variety of other "unspecified" substances. The issue of security leaks at the Calverton hangar and the criminal possession of classified government evidence further served to obscure the value of that evidence. Finally, federal authorities stated that the residue was determined to be inconsistent with rocket propellant, and thus of little to no value in answering the question of what caused the disaster.

If this were the case, why dispatch agents to investigate my home, track down my media contact, then demand and receive one tiny bit of material that federal experts say is of no consequence? How significant is this one tiny bit of cloth, when in Calverton hangar and in undisclosed locations, federal investigators have access to large supplies of the same material?

There can be no other explanation: The U.S. government knows that TWA Flight 800 was shot out of the sky by accident by an unarmed missile launched during a military exercise—but is attempting to hide this fact from the families of the victims and from the American people.

APPENDIX I

Navy FACSFAC Indicating Military Activity in Zone-5 on July 17, 1996

```
**************************
* U N C L A S S I F I E D *
**************************
```

PATUZYUW RUEOMCC0750 1981914-UUUU--RUEDJKA
ZNR UUUUU
ZUI RUCOMAR0328 1981807
P 161887Z JUL 96 ZYB
FM FACSFAC VACAPES OCEANA VA//OAC//
TO RUEDJKA/FAA NEW YORK ARTCC RONKONKOMA NY//MOS//
BT
UNCLAS //N03722//
MSGID/GENADMIN/FACSFAC VACAPES//
SUBJ/ALTRV APREQ TANGO-BILLY/7//
REF/A/TEL/FFVC/16JUL96//
AMPN/PHONCON BTWN FFVC OAC/MS COSBY AND FAA ZNY/MR BOMBROWSKI//
POC/S. COSBY/FFVC OAC/FFVC/-/TEL:DSN 433-1233
/TEL:COMM 804-433-1233//
RMKS/1. REQUEST STATIONARY ALTRV TANGO-BILLY SFC TO FL100 WITHIN
3950N 7210W, 3950N 7045W, 38N 7045W, 38N 7210W (EXCLUDING WARNING
AREAS) FROM 180100Z TO 180700Z JUL 96.
2. ALTRV APVD BY REF A, NO REPLY REQUIRED.//
BT
#0750

DELIVER TO:

177

APPENDIX II

CLASSIFIED NTSB CHAIRMAN'S BRIEFING/STATUS REPORT: NOVEMBER 15, 1996

TWA 800
Chairman's Briefing/Status Report
November 15, 1996

The Chairman opened the meeting by thanking all the personnel involved with the investigation of TWA 800. He stated that this investigation is the largest investigation ever taken on by the Safety Board and will dictate how investigations will be run for many years in the future.

The Chairman indicated that with the new Congress about to start there are many new people in office and feelings may be changing about the way they look at the TWA 800 investigation. Hearings may take place concerning the investigation as soon as February 1997 and the Safety Board should be in a position in which as little criticism as possible can be levied against it. Therefore we should be continuing on in every aspect of the investigation with the constant goal of finding out the cause of the accident. The Chairman emphasized that all involved must continue to work effectively, efficiently, and to communicate clearly. We should anticipate the need to defend every decision that we make.

Investigators were asked to provide projections of future investigative work items, completion dates, and completion of factual reports. They were asked to project their work based on the assumption that there was no FBI involvement and work would proceed according our NTSB established procedures.

Forensic Pathology / Medical Examiners Group—Burt Simon

Field note preparation now in progress to include mapping of medical information for graphical presentation and analysis. Dr Shanahan will be at the Calverton facility until November 18, 1996 analyzing and comparing medical and cabin data. Although Burt Simon agreed to complete his factual report by 1/15/97, an addendum may be required if more human remains are recovered during the continued trawling operations. Burt will include toxicology findings in his factual report.

Cabin Furnishings Documentation—Hank Hughes

181

Completion of the data base, quality control, and analysis is about 80% complete. Construction of the cabin interior will continue as new parts arrive from the trawling operations. Integration of digital photographs with the data base is about 70% complete. Completion of the center fuel tank foam-core model is 60% complete.

Hank Hughes agreed to finish his factual by 1/1/97. Depending on new wreckage recovered by the trawling operation, he has about 2 to 3 weeks on scene work left to accomplish.

Search and Rescue—Matt McCormick

The initial search and rescue will be documented and a report produced by a person to be assigned by the chief of the Survival Factors Division. A time line will be developed concerning notification, response, vessels, aircraft, etc. Copies of reports will be obtained from all the agencies involved.

Since the data are already available from Coast Guard and Navy documents, a completion date of 1/1/97 should not be a problem.

Data Base Management—David Mayer/Debbie Bruce

The group will continue to manage and refine the charting/plotting documentation with Oceaneering and the part tagging project especially since the trawling operation started. Oceaneering should have the data base complete about 2 weeks after the trawling operations have been completed. The Safety Board should maintain a CAD capability after Oceaneering is released from the Navy contract (contract support is underway).

The means to scan field notes, factual reports, and other documentation for party distribution and docket preparation are underway. The Director of Research and Engineering directed that the scanning in of documents will take place in Washington. The Systems Group Field notes have been completed and brought to Washington and will be scanned in by the end of next week, 11/22/96. They will be followed by the Medical Examiner's and the Cabin Documentation notes and should proceed on schedule with other groups' activities.

The computer modeling project, CATIA, was discussed. The Chairman indicated that we should collect proposals to model at least the center fuel tank. Debbie Bruce will define the project and collect bid estimates for it. It was pointed out that this software is primarily used as a demonstration aid rather than an analytical tool.

ATC—Al Lebo

The field factual was completed and distributed approximately 10 days after the accident. Al reported no safety issues for ATC. TWA flight 800 was a "Life Guard" flight (requiring priority ATC handling) due to the medical tissue it was carrying. Al noted that there was a 40 second gap between the last recorded radar hit and the time the explosion was mentioned on any ATC tapes. He has reviewed the full FAA ATC package with requested transcripts. We can authorize the FAA to release the ATC tapes to the media.

The Chairman requested that we obtain all the ATC air/ground tapes 15 minutes prior to the accident that referenced anything to do with TWA flight 800 and that Al should personally audition all these tapes. Every word on those tapes must be agreed on by the group as a whole.

Al Lebo agreed to get the DEA to confirm a statement in his field notes that indicates that the FBI followed up on this item. He has delayed the completion his factual report till 1/31/97.

Radar Data—Charlie Pereira

The Chairman indicated that Charlie may also want to listen in on the ATC tapes. Ron Schleede indicated that the FBI and the DIA have indicated to him that there is a possibility that Geo-Synchronous satellite sensors may have picked up either radio transmissions or infrared emissions related to this accident. Charlie will request to the DOD in writing any radar data that might pertain to this investigation. The Chairman requested that Charlie Pereira personally review all radar data concerning TWA flight 800 and that all tapes will be maintained here at the Safety Boards headquarters. Charlie will submit his report by 12/31/96.

Ron Schleede will write a letter for Bernie Loeb's signature to Ron Morgan for a full explanation of the FAA handling of ATC and radar tapes concerning TWA flight 800. The letter will reference the technician who did the analysis resulting in conflicting radar tracks that indicated a missile. It will also inquire why that information was reported to the White House and sent to the FAA Technical Center before the Safety Board was given access to the data.

Witness Group—Norm Wiemever

Due to the initial information that this investigation would at any moment turn into a criminal investigation, Norm only had access to witness statements and was not able to take notes or prepare summaries of the interviews. Redacted statements prepared by FBI agents for ground personnel involved with the dispatch of TWA 800 have recently been provided to our team; however, a review indicated that the proper accident investigation questions had not been asked.

At our request the FBI handed over 5 volumes of witness interviews (302 forms) to the Witness Interview group. A review of these statements is currently going on and a schedule is being set up to re-interview many of these people. The Chairman directed that if there is any resistance from the FBI we are to get it in writing. Input will be provided to other groups which may result in further interviews. A court reporter will be hired to speed up the recording process. Pam McKenzie will coordinate the hiring.

Norm indicated that the witness group field notes would not be ready until 2/15/97.

Operations/Witnesses—Norm Wiemever

Norm has found no safety issues in the operations area. He needs to fill in areas concerning fueling, dispatch release, weather, weight and balance and aircraft cockpit procedures. His group will reconvene to review the CVR transcript for procedures followed. Norm emphasized his work also included witness interviewing. Norm will incorporate some of the information as necessary from the witness interviews. This will most likely delay his operations factual until 1/15/97.

Powerplants—Jim Hookev

The powerplants group chairman was not present at the meeting but Al Dickinson reported that the engine examinations have been completed with no safety issues apparent. Jim had agreed to complete his factual report by 1/1/97. The delay involves other multiple investigative tasks that include travel.

Jim had not been able to obtain the photographs taken by his group during the engine examinations and given to the FBI for processing. He agreed to return to the hangar to obtain the necessary photographs for his factual report. Al Dickinson has obtained all the photographs which are available to Jim whenever he can return to look at new engine parts recovered.

Although Jim does not see any need to retain the engines or accessories for NTSB purposes, a survey of the party coordinators and other group chairman revealed that nobody was in favor of releasing any parts until a cause for the accident has been determined. The Chairman agreed that no parts should be released at this time.

Airport Security—Larry Roman

Larry did not attend the meeting, however, Matt McCormick indicated that his field notes are complete. Group members do not have copies of notes because of sensitive information in them. A meeting will take place with Dan Cambell, Tom Lasseigne, and GC representatives from the FAA and the FBI in order to resolve what can be released to the public without compromising security. Larry's factual report will be completed by 12/15/96.

Airport Security Factors: Explosive Detection Canine Program—Tom Lasseigne

Tom prepared the draft report on 9/24/96 regarding the explosive spill on June 10, 1996 at the St. Louis-Lambert Airport. He interviewed police and FAA airport, canine program , and explosive division personnel. He has identified gaps in the explosive handling program which may be indicative of system wide deficiencies. Tom anticipates the preparation of a factual report for the public docket by 12/20/96.

185

Maintenance Records—Debra Eckrote

The maintenance log from the airplane was not recovered and recent sheets may have not been pulled. The practice is not a violation because the PMI accepts the practice. Further investigation may result in a recommendation in this area.

Debbie and other team members discussed several open questions that need review in the records, including items pertaining to possible previous fuel leaks, fueling problems with the accident airplane, and maintenance history on the fuel probes. Debbie also discussed her observations during meetings at Boeing at which fuel/air explosion modeling took place. Debbie might have to travel to Kansas City at some point to verify some information in the records. She will be reviewing new statements from the witness group as necessary. She agreed to complete her factual report by 1/1/97.

The Director of RE indicated that he would check with Bob MacIntosh concerning any data that Air Claims would have concerning this aircraft.

Metallurgy—Frank Zakar/Jim Wildev/Mike Marx/Joe Epperson

The Materials Laboratory has been supporting the structures, systems, and fire and explosion groups in the hangar and has prepared reports of examinations as they are completed. Some positive findings included through cracks found in some titanium bleed air ducts that may have resulted in hot air leaking around them. Many of the ducts, however, could not be identified due to a lack of part numbers on the pieces being examined. RE-30 will assist Deepak in re-examining parts that Boeing has raised questions about. Additional items are in the lab and reports will be completed shortly on those examinations.

No evidence was found of fatigue in fuselage structure for pieces associated with the "red zone" A discussion was also held regarding some areas of fatigue cracking found in the forward spar, and possibly on the aft spar, areas of the CWT. Fragment evidence found in the CWT was indicative of an explosive event, not necessarily a missile or a bomb but rather associated with the CWT explosion.

186

It was agreed that a group of metallurgists would return to Calverton and work with the structures group to go over the probable failure sequence and reexamine the center wing tank fractures. This group would be reconvening right after the Thanksgiving vacation. The chairman indicated that he wants us to find outside experts to review our facts and conclusions. Mike Marx will coordinate obtaining these experts for the metallurgical effort.

Lab personnel should not have difficulty completing any reports of examinations requested by other groups by 1/1/97.

Three-Dimensional Reconstruction Project—Larry Jackson

Larry has prepared the contract for a Three-Dimensional Reconstruction Project and is ready to proceed with the project. At the request of AD the contract was revised to require written approval prior to initiating each task. Although the Chairman has given his go-ahead, Dan Cambell agreed to review the contract before it is approved. It will also be reviewed by the Managing Director. The project is scheduled to be initiated by 11/22/96.

All NTSB staff agreed that the project was not necessary for our investigative work. They agreed that it would be a good illustration; however, it would not enable us to find the cause. The existing mockup of the CWT and center fuselage is considered sufficient to understand the accident.

At the request of the Chairman, AD will assign someone who understands government surplus property to look into wreckage storage property closer to Washington and less expensive. Paul Voorhees stated he would handle this project. This would involve moving all the wreckage but may result in a cost saving in the long run.

Trawling Project—Charlie Pereria

The project was started on 11/3/96 with 2 ships in area 1 (green area). The recovery has been very successful and although the area has been reduced in size, one ship is continuing to find parts. The other ship has been assigned to a triangular area between the green and red areas. Two additional

187

ships arrived on 11/11/96 and have been trawling areas 2 and 3 (the red area with the yellow area inside of it).

The trawling project continues to find pieces of wreckage in all areas and the Chairman indicated that it should continue until the contract money has been expended (approximately 18 days from 11/14/96). The Chairman requested that a map be produced to indicate what areas have been trawled and what areas will be trawled. Charlie stated that Oceaneering keeps that data and updates it daily and he would make sure it would be supplied to the Chairman.

Aircraft Performance—Charlie Pereira

Review of the FDR data has not indicated any anomalies. Boeing has compared the data from the accident flight with other comparable flights and found no significant deviations from normal flights. John Clark is comfortable with all of this Boeing data. Charlie indicated that his factual report will be completed by 1/15/97.

CVR—Jim Cash

Report completed by 12/31/96.

Spectrum Analysis—Jim Cash

Report completed by 12/31/96.

FDR—Dennis Grossi

Report completed by 12/15/96.

Trajectory Study—Dennis Crider

The completed work consists of documenting fuselage skin and center fuel tank parts found in the red area up to 10/15/96, A/C pack pieces and the forward keel beam, and selected cabin interior parts. The report needs to incorporate wreckage diagram changes as part positions are finalized and add new wreckage items as they come in.

The Chairman requested that the report also include weather data and sea conditions and their effect on the location of pieces found. Dennis agreed to include this information in his report and to complete his factual report by 1/30/97.

Structures—Deepak Joshi

Currently approximately 15% of the center wing tank, 12% of the fuselage (skin), 15% of the left wing, and 8% of the right wing is missing.

Deepak indicated that the new wreckage being brought in as a result of the trawling operation has slowed down his documentation efforts. A discussion ensued concerning obtaining additional help. The Chairman stated that he would be glad to call the top men in each of the organizations but Deepak indicated that he did not think that was necessary at this time. The result was that Deepak does not want any new people because it would take too long to train them. He indicated that when he returns from his trip to observe other reconstruction projects he will coordinate with all the parties and come up with a work plan to optimize the use of experienced people in each of the parties. He will attempt to get the FAA members on his group to work on weekends.

He agreed to complete his factual report by 2/15/97.

Fire and Explosion—Merritt Birky

Merritt indicated that to date, his group has seen no evidence of erosion or pitting in any of the wreckage. He also indicated that they have not discovered any static or fuel transfer problems with the center wing tank. Merritt will be hiring a fuel explosive expert from Cal Tech as a consultant as a continuing effort to bring new ideas into the investigation.

His group has extensive future work off scene pertaining to explosion testing and modeling of explosions. A discussion about purchasing old 747's followed but it was concluded that it was to early too start the process since it had been looked into back in August and it was found that most owners want commitments within a few weeks.

189

Merritt agreed to complete his on scene work and factual report by 1/1/97.

Systems—Bob Swaim

Bob stated that he brought his systems field notes with him for scanning. He indicated that he has narrowed down his systems examinations to 5 fuel energy sources. He stated that Honeywell is conducting a FMEA on the entire fuel system. He indicated that he would appreciate any comments on his recommendation letter as soon as possible. He also indicated concern about the upcoming Boeing temperature profile test scheduled for 12/16/96. He will be contacting Boeing to obtain thermocouple data that will be generated and recorded during testing prior to the 16th.

Bob agreed to complete his on scene work and factual report by 1/1/97

SUMMARY INFORMATION

The Chairman discussed the hiring of an editor to document everything that has taken place since the start of the investigation. Further coordination will be needed to insure that takes place

NTSB staff will be done with the on scene (hangar) documentation by 2:15/97 They will have their factual reports completed by that time, with some exceptions that do not require visits to the hangar. Therefore, it would be conceivable that we would not require access to the wreckage after that time if the reconstruction project does not proceed.

If the "second set of eyes" concept is pursued, it should be done after the NTSB group chairmen complete their work. It is projected that that work would take about 1 month. Therefore, the "second set of eyes" effort would be completed by 3/15/97.

APPENDIX III

NTSB CONTROLLED METALLURGY STRUCTURES SEQUENCING GROUP REPORT, 1/24/97

Group Description

The Metallurgy and Structures Sequencing Group was formed to develop
scenarios for the sequence of structural breakup of the airplane and to correlate proposed
scenarios with the structural observations. The primary focus of this report is to address
the wing center section (WCS) breakup sequence and any potential interaction or
relationship with the fuselage "red area" breakup sequence[1]. In addition the report will
address overall airplane breakup sequence in somewhat less detail. The Group examined
the airplane structure from December 2, 1996 to December 13, 1996, and from January 7,
1997 to January 22, 1997.

1.5   List of Abbreviations

    WCS   Wing Center Section
    SWB   Spanwise Beam
    STA   Fuselage Station
    BL   Buttock Line (lateral distance from centerline of airplane)
    RBL   Right Buttock Line
    LBL   Left Buttock Line
    RHS   Right Hand Side
    LHS   Left Hand Side

---

[1] See Structures Group Notes for further description of the recovery areas.

The basic narrative of the main part of this document is intended to represent a summary report. A more detailed rationale for the sequence elements may be found in Appendix B for most components. The figures referred to in the basic narrative are found in Appendix A. This document refers to sooting patterns, fire damage, and structural damage and description throughout. Limited sooting diagrams and structural diagrams are provided in figures and appendices A and B. For a more detailed accounting of these features refer to the Fire and Explosion Group and Structures Group documentation. Some identified discrepancies between the Sequencing Group's observations and the Fire and Explosion Group notes were referred to the Fire and Explosion Group for resolution.

3.2   Description of Wing Center Section

The wing center section (WCS) is a large box with an airfoil shape generally corresponding to the shape of the inboard wing. The WCS is bounded by the wing front spar, wing rear spar, side-of-body ribs, and upper and lower panels. Spanwise beams #1, #2, #3, and the midspar form intermediate inboard-outboard beams. There is a fore and aft beam at the airplane centerline between the rear spar and the midspar. Most of the internal volume of the WCS, the volume between the rear spar and spanwise beam #3, forms the center fuel tank on a 747-100. The remaining volume of the WCS, between spanwise beam #3 and front spar, is a dry bay and does not contain fuel in the 747-100. See figure 3-1 for a WCS schematic.

The Group examined wreckage that had been recovered and identified from the WCS in three separate reconstruction mock-ups. The WCS upper panel, rear spar, spanwise beam #1, midspar, centerline rib, spanwise beam #2, and spanwise beam #3 formed one reconstruction. The front spar, forward most lower panel pieces, keel beam, and adjacent fuselage pieces from the red area (minus the upper lobe pieces) formed a second reconstruction. Finally the remaining WCS lower panel was reconstructed separately.

The upper and lower panels were more than 95% recovered and identified. Recovery and identification of other major components ranges from 95% to approximately 65% on spanwise beam #2 and less than 30% on the left side-of-body rib.

Approximately 70% of the front spar, 60% of spanwise beam #3 and the manufacturing access door from spanwise beam #2 were recovered from the red area indicating relatively early departure from the airplane.

3.3   Description of the Fuselage Red Area Pieces

The fuselage pieces recovered from the red area are enveloped between the wing front spar at fuselage station (STA) 1000 and STA 741. The fuselage red area

194

pieces were examined in two separate reconstructions. Those generally below the main deck window level were included in the reconstruction mockup with the wing front spar bulkhead and pieces of fuselage from the green area. Upper lobe fuselage red area pieces were laid out on the floor relative to each other. The basic fuselage skin on the pieces recovered from the red area is more than 95% recovered. The discussion of the sequence for the fuselage red area pieces is contained in section 6.9.

4.0   WING CENTER SECTION SEQUENCE

4.1   Upper WCS Panel Sequence

The upper skin panel of the WCS is more than 95% recovered and identified with several missing areas on the far left side and two small missing areas in the middle. All identified pieces of the upper panel were found in the green area. However, there are dramatic differences between the left side (clean) and right side (sooted) on both the top and bottom surfaces of this panel. A close examination of sooting on both surfaces and mating fracture faces yields a definition (see figure 4-1) of material departing with the left wing (minimal sooting) versus right wing (heavy sooting) at the time of major airplane breakup. There is an area of upper panel material between the left side of body and approximately LBL 34 that either separated independently during wing breakup or remained attached to the right wing for a time after major airplane breakup.

The reconstructed upper panel showed a multiple wave shape, indicative of spanwise compression buckling. In addition, the longitudinal fractures in the upper panel are generally typical of bending (buckling) overstress separations. These fractures and the compression buckling are indications of up bending loads on the wings at "G" levels beyond the structural capability. Stress analysis would also indicate that early loss of the front spar and spanwise beam #3 would significantly weaken the ability of the more forward upper panel to carry compression loads but would not initiate overall panel collapse under nominal flight loads.

4.2   Lower WCS Panel Sequence

The lower skin panel is more than 95% recovered and identified with small missing pieces on the left side and right middle area. The sooting patterns on the lower surface of this panel varies from light to heavy in different areas over essentially the entire lower surface, but with the heaviest accumulation of soot on the right side of the lower surface. The upper surface shows more localized areas of heavy sooting with some areas clean. The soot patterns on the upper and lower surfaces and on the fracture faces also indicate a delineation between material separating with the left wing versus the right wing as shown in figure 4-2. Fracture features along this line of delineation are typical of a tensile overstress, also consistent with wing up bending. The lower panel fracture appears to have started between SWB#1 and the midspar (at left side-of-body).

In addition to the presence of heavy soot accumulation associated with a major fire after major airplane breakup, there are two additional sooting patterns that suggest prior fire sources: (1) Sooting on the lower surface of the lower panel, including some heavy sooting adjacent to the left side-of-body, and (2) heavy sooting on the right portion of the upper surface of piece CW221 (generally between SWB#2 and the front spar, and right of BL0).

The right side-of-body rib is more than 75% recovered and identified with a number of small pieces which cannot be accurately placed in the reconstruction. The rib has moderate to heavy sooting on the inboard surface of the areas between SWB#2 and the front spar. The rib stayed with the right wing on major airplane breakup with most fractures probably occurring on water impact.

4.4   Left Side-of-Body Rib Sequence

Only a small percentage (less than 30%) of the left side-of-body rib has been recovered and identified, essentially all of which is between the rear spar and SWB#2. Identified pieces are broken into small fragments with negligible sooting. The lack of sooting indicates that the recovered and identified portions of the left side-of-body rib stayed with the left wing following major airplane breakup. Breakup of the rib into a large number of fragments is consistent with water impact, similar to the fragmentation that occurred to the left inboard upper wing skin (see section 8.1). Both the side-of-body ribs and the wing upper skin are comprised of 7075 aluminum alloys with characteristic high strength and relatively low elongation properties compared to the lower skin.

4.5   Rear Spar Sequence

The rear spar is approximately 90% recovered and identified with missing pieces mostly on the left side (LBL 57 to LBL 98) and a small area on the right (RBL 22 to RBL 33). Both the forward and aft surfaces of the rear spar are sooted to the right of LBL 21 (very heavily between LBL 21 and RBL 63). A review of sooting, fracture morphology, and interface with upper and lower panels indicates that at major airplane breakup, the spar generally to the left of LBL 21.5 (CW1006 and CW1007) departed with the left wing, and the region to the right of LBL 21.5 departed with the right wing (see figure 4-3). No identifiable indications of damage or sooting prior to major airplane breakup could be documented.

4.6   Spanwise Beam #1 Sequence

Spanwise beam #1 (SWB#1) is approximately 90% recovered and identified with the majority of missing material on the right side and the remainder mostly distributed full span across the lower portion of the beam. Sooting varies from clean to heavy on both the forward and aft surfaces of SWB#1 with a number of mating fracture faces equally sooted. A review of sooting patterns, electrical conductivity readings, and crack morphology indicates there were likely multiple failures at the time of major airplane breakup (RBL 66 to LBL 57) with most of the beam (almost to the left side-of-body) going with the right wing (similar to lower panel, see figure 4-4).

The access doors on both sides of centerline have consistent edge band deformations between fasteners and consistent patterns of soot moving aft through the openings. This is indicative of an earlier event forward of SWB#1 involving overpressure while the wing center section was still relatively intact. The sooting is indicative of the presence of sustained fire and soot following initial overpressure and preceeding major airplane breakup. Deformations and soot patterns on the left side door are more pronounced than on the right side door. This lack of uniformity suggests that the centerline rib between SWB#1 and midspar may have been at least partially present when an overpressure event occurred.

4.7   Midspar Sequence

The midspar is approximately 75% recovered and identified with areas missing on both left and right sides. Sooting (light to heavy) is generally present on the forward surface (RBL 67 to LBL 44) and aft surface (RBL 67 to LBL 98). Sooting of mating fracture faces, and crack morphology indicate that the midspar failed at LBL 44 during major airplane breakup with the area to right going with right wing and remainder with left wing (see figure 4-5). Relatively minor sooting outboard of LBL 44 on the aft surface is another indication of an earlier event involving fire/soot between SWB#1 and midspar (see discussion on SWB#1, section 4.6). The midspar did not contain indications of differential pressure between the forward and aft sides.

4.8   Centerline Rib (BL 0.00 Rib)

Approximately 90% of the centerline rib between the rear spar and SWB#1 has been recovered and identified but only 40% of the rib between SWB#1 and midspar (see figure 4-6). The section between rear spar and SWB#1 is more heavily sooted on forward, aft, and upper fracture faces. The pieces between SWB#1 and midspar are equally heavily sooted on both the left and right surfaces and most fracture faces. Sooting and the location and features of fractures indicate the centerline rib remained with the right wing at major airplane breakup. Definitive damage or sooting prior to major airplane breakup could not be identified, however see the sections on SWB#1 (section 4.6) and the midspar (section 4.7) for discussion of indications of earlier damage on these components, which may have also related to the centerline rib. There were indications described of an overpressure acting aft on SWB#1 and early presence of fire or soot ahead of SWB#1, either of which might have affected the centerline rib.

4.9   Spanwise Beam #2

Spanwise beam #2 (SWB#2) is approximately 65% recovered and identified with most of the left side still missing (see figure 4-7). The manufacturing access door and a small attached portion of web above the door were recovered from the red area indicating early departure from the airplane. The door fasteners on the bottom and left (inboard) sides of the door were separated mostly in vertical shear (door along with upper and outboard surround structure moving up and the remaining surround structure moving down relative to each other). The remainder of the door fasteners were fractured in tension by the door peeling forward and upward, finally tearing out a small portion of the upper web above the door. Witness marks found on the upper panel corresponded to deformation in the lower inboard corner of the door, indicating that the door separated upward with enough velocity to create this damage. Final separation of the door (peeling upward in the forward direction) indicated that the pressure on the aft surface of the door was significantly greater than the pressure on the forward surface of the door at that time. The access door is only lightly sooted, while sooting is moderate to heavy over most of the other pieces of the beam, consistent with much more substantial fire exposure after separation of the access door. The soot patterns indicated that most of the identified pieces of SWB#2 (with the significant exception of the access door) remained attached to the right wing at major airplane breakup.

A large portion of the right side of SWB#2 remained attached to the upper skin panel. Soot patterns indicate that the lower chord remained attached to the web until water impact but was separated from the lower skin panel before major fire exposure. In general, the features on the right side of the beam indicated that this entire portion of the beam remained largely intact but had separated from the lower panel before fire exposure. Recontact damage and separation of the web from the lower chord occurred after fire exposure.

The right side of SWB#2 also contained "accordion" damage (folding directly inboard) from forces acting in the inboard direction on the outboard end of the beam. No soot accumulation occurred after the deformation was created (see appendix B).

Close attention was directed to the keel beam interface (see Appendix B) where fracture of the two major tension bolts was due to a tensile overload (consistent with downward motion of the forward piece of keel beam as described in section 5.1). Early events associated with SWB#2 included the previously discussed initial separation of the access door surround structure in shear and tensile separation of the fasteners common to the SWB#2 lower chord and lower panel. These features could be consistent with either a large downward load imparted by the keel beam tension bolts or overpressure acting approximately in equal amounts in the bays ahead of and behind SWB#2.

4.10   Spanwise Beam #3

Spanwise beam #3 (SWB#3) is approximately 85% recovered with most of the missing area located between RBL 50 and RBL 90 (see figure 4-8). Pieces of the beam between approximately RBL 50 and LBL 80 were recovered from the red area indicating relatively early departure from the airplane. The part of SWB#3 from RBL 87 to right side-of-body is heavily sooted (burn damage) on both surfaces as well as most fracture faces. The pieces recovered from the red area are generally lightly sooted on both faces with occasionally more sooting on the front face. The pieces between LBL 80 and the left side-of-body (green area) are clean on the front and moderately sooted on the aft surface. During major airplane breakup the pieces outboard of LBL 80 went with left wing and the pieces outboard of RBL 87 with right wing. The sooting on the aft surface on the left side and on red area pieces is indicative of an earlier event. Larger amounts of soot accumulation and fire damage on the right wing pieces indicate that this portion of the structure was involved in a later major fire. The soot and fire damage associated with the later major fire masked any possible features that may have been associated with a possible earlier fire affecting the pieces of SWB#3 to the right of RBL 57.5.

SWB#3 contains vertical stiffeners on the aft face of the web. Approximately every third stiffener is attached to the upper skin at a (forward to aft) floor beam location over the upper skin panel of the WCS. The upper chord of SWB#3 has a "Z" shape, with the upper horizontal leg pointing aft.

The upper chord for SWB#3 was fractured through the fillet radius between the vertical leg and the upper horizontal leg of the chord. The upper chord fracture initiated at multiple locations and progressed essentially the full width of the beam.

Witness marks and deformation associated with separated fasteners for the stiffener fittings at the top of the beam were indicative of both an upper motion of the upper skin panel and a forward motion of the upper portion of the beam as this area separated. Initial separation of the upper portion of SWB#3 was consistent with overpressure on the aft face of the beam, causing the upper panel of the WCS to move upward a small distance as the upper portion of the beam rotated forward.

Vertical fractures through SWB#3 were found at various locations including near the respective sides-of-body. These fractures are also consistent with a forward acting overpressure on the aft surface of SWB#3. Separation of the top of the SWB#3 allowed the segments to rotate forward about the lower intercostals (or, in the case of the center segment, about the beam's lower chord) until the top of the beam impacted the stiffeners on the aft surface of the front spar approximately 12" below the upper skin panel. The impact broke off parts of the upper web and stiffeners of SWB#3. The remaining lower portion of SWB#3 continued rotating forward and down with upper stiffener ends tearing vertical holes in the front spar web at various locations down to about 1 to 2 feet from the lower panel.

Following separation of the upper end of SWB#3 from the upper panel of the WCS, the forward rotation of the upper end of SWB#3 resulted in tension fitting separation at the interface with the keel beam (see appendix B, page SWB3-3 and associated figures B-6 and B-7) without separating the tension bolts for these fittings. Separation of the fittings at this time resulted in free play of about 0.65 inch in the bolts. Downward movement of the keel beam (later in the sequence, as described in section 5.1) took up the free play and separated these tension bolts.

4.11   Front Spar Sequence

The front spar is more than 95% recovered and identified (see figure 4-9). Pieces of the front spar between approximately RBL 50 and LBL 110 were recovered from the red area indicating they departed the airplane as part of a relatively early event. One piece, CW 504 from left side was recovered to the west of all other major structure in the red area. There is localized heavy sooting on the forward surface of the lower right portion of the front spar outboard of the wing leading edge vapor seal rib and around the dry bay access opening, primarily below and outboard of the ring chord. The pieces of the front spar that were recovered from the red area have minimal sooting. The front spar outboard of RBL 50 went with the right wing during major airplane breakup while that outboard of LBL 110 went with the left wing.

As discussed in section 4.10, SWB#3 rotated forward impacting the vertical stiffeners on the aft surface of the front spar. The impact, along with possible overpressure from behind SWB#3 fractured the front spar upper chord in the radius between the horizontal and vertical legs of the chord. The horizontal leg of the chord remained attached to the upper skin panel, and the vertical leg remained attached to the web of the front spar. Continued forward and downward rotation of SWB#3 tore holes in the front spar web, at various locations down to about 1 to 2 feet from the lower panel. Geometric layouts (see Appendix B) indicate that SWB#3 probably rotated almost fully forward and down prior to significant rotation of the front spar about its connection to the WCS lower panel. The potable water bottles (centered on the front surface of the front spar) sustained relatively minor damage on their aft sides from impact with pieces of SWB#3. The forward sides of the bottles did not display obvious impact marks from contact with the cargo floor structure. A geometric layout indicates that only about 10 degrees of rotation of the front spar would be needed to force contact between the bottles and the cargo floor structure. The lack of damage to the forward sides of the bottles indicates that the front spar rotated less than 10 degrees before the cargo floor structure had begun departing the airplane (see figure 4-10).

Forward rotation of the front spar about its lower end is consistent with overpressure loads released by SWB#3 as it rotated forward. Deformations in the upper chord of the front spar were in a symmetric "sine wave" shape, with a lobe bulged forward on each side of center. Forward deflection amplitude was maximum at approximately LBL 66 and RBL 66 and minimum at the approximate center of the span of the spar (corresponding to the potable water bottle locations). Tension separations of the vertical leg of the front spar upper chord were found in multiple locations (LBL 66 and RBL 48, see figure B-12 in appendix B) corresponding to the forward bulges on each side of center. These separations are consistent with tension generated by the stretching of the vertical leg of the upper chord as the upper portion of the front spar rotated forward. Vertical fractures through the front spar web progressed downward to the lower pressure bulkhead. Compression buckling of the vertical stiffeners attaching the front spar to the lower pressure bulkhead (located below the front spar and above the ring chord, see figure 4-10) indicates that separation of the front spar pieces from the lower skin panel and the lower pressure bulkhead was as a result of forward rotation of the front spar pieces about the lower chord caused by impact loads and/or pressure loads on the aft surface of the spar.

The front spar is attached to the keel beam through four bolts (5/16" diameter) through fittings on the aft edge of the front spar stiffeners above the keel beam. Tension separation of these bolts is consistent with the forward rotation of the front spar.

Close examination revealed small pre-existing fatigue cracking areas in the upper and lower shear ties for the stiffeners on the aft surface of the front spar, in the front spar lower chord near the underwing longerons, and in a longitudinal floor beam detail. The shear tie fatigue cracks and front spar lower chord fatigue cracks are in areas subject to Service Bulletin directed inspections and/or modification. There are no indications that the fatigue cracks contributed to initiation of the breakup sequence or affected the subsequent breakup pattern. See Appendix C for a further discussion on pre-existing fatigue cracking.

4.12 .   Front Spar Lower Pressure Bulkhead and Local Interface Sequence

The front spar lower pressure bulkhead is an extension of the plane of the basic WCS front spar downward to the fuselage skin which starts at the bulkhead and extends forward. The lower bulkhead is bounded on left and right sides by the underwing longerons and associated fittings. The lower bulkhead web is spliced to the main WCS front spar web just below the front spar lower chord and joined to the fuselage skin by an angle "ring chord". The splice between the webs of the lower bulkhead and the front spar is reinforced by vertical stiffeners on the forward side which effectively form an extension of the upper WCS front spar web stiffeners on the aft side. The lower bulkhead is also directly connected to the keel beam at LBL 9 and RBL 9.

The lower pressure bulkhead has been essentially 100% recovered, and pieces between LBL 66 and RBL 66 are either from the red area or are unconfirmed. It is noteworthy that on both sides the portion outboard of BL 66 associated with the underwing longeron and adjacent fittings stayed with the airplane and were recovered from the green area. Sooting was negligible on the lower bulkhead. Crack propagation directions have been identified and documented on Figure 4-11.

At RBL 66, LBL 26, and RBL 66 web cracks propagated down from the front spar web and reinitiated downward in the lower pressure bulkhead with eventual associated axial fracture of the ring chord. There are additional lower bulkhead fractures at RBL 9 and LBL 9 which are close to the keel beam interface. There is an additional vertical web crack at LBL 49 which is associated with the departure of piece LF55. The stiffeners "splicing" webs of the front spar and lower bulkhead are uniformly buckled in the free flange consistent with the motion of the front spar rotating forward. The stiffener at LBL18 is not bent forward as far as the others indicating limited forward rotation of the front spar in this area prior to ring chord separation at the bottom of the stiffener. The fasteners common to the splice between the webs of the front spar and lower pressure bulkhead are consistently (left and right sides, BL26 to BL75) separated in shear with the lower web being pulled downward and somewhat inboard.

There are two bathtub fittings nested in the ring chord above the underwing longeron. These joints have fractured in a tension/bending mode consistent with the fuselage skin panels forward rotating outward about the ring chord, applying a moment which is reacted between these fittings (tension) and the longeron fitting. The bathtub fittings appear to have failed first then the longeron joint in a manner consistent with being overloaded by the same bending moment.

The keel beam lower chords are spliced just ahead of the lower pressure bulkhead to the keel beam runout in the forward body. Each keel chord extension tang is fractured identically in a down bending mode (i.e. body panel with keel runouts rotating downward relative to the main keel beam aft of the front spar).

The integration of significant lower bulkhead fractures into the overall sequence is accomplished in Section 7.0.

5.0   KEEL BEAM AND OVERALL WING CENTER SECTION SEQUENCE

5.1   Keel Beam Sequence

The keel beam (see figure 5-1) is located along the centerline of the airplane under the WCS from below the front spar aft through the wheel wells to the STA 1480 bulkhead. The beam is a box structure with two vertical webs (at LBL 9 and RBL 9). Each web has a heavy chord along its lower edge and a smaller chord along its upper edge. The upper chord is attached to the lower surface of the lower skin panel of the WCS through a series of aluminum rivets forward of the midspar and titanium bolts aft of the midspar as well as stronger steel tension bolts at each transverse beam inside the WCS (front spar, SWB #3, SWB #2, midspar, SWB #1, and rear spar).

Almost the entire keel beam has been recovered and identified. The forward 13.5 feet of the beam (from the front spar to 22 inches aft of the midspar) separated from the remainder of the beam. The forward portion of the beam contains no confirmed sooting (as of the date of this report), and was recovered in the red area (indicating early departure from the airplane). The aft portion of the keel beam was recovered from the green area, and this section of the beam contained moderate to heavy sooting, indicating that it remained with the right wing for a period of time following major airplane breakup. The aft section of the keel beam also separated 1) just forward of the STA 1350 bulkhead and 2) along the upper flange where it had been attached to the WCS lower panel. These fracture areas did not contain soot, indicating that the aft portion of the keel beam separated from adjacent structure after sooting conditions ceased, probably at water impact.

The forward keel beam piece separated from the aft piece with a similar fracture through the web and chord on each side of the beam. The web fractures progressed from the top of the webs to the bottom, consistent with a downward bending moment on the keel beam. The large chords at the bottom of the beam webs also fractured in downward bending (forward end of the beam moving down). Separation of the upper edges of the keel beam from the lower skin panel of the WCS involved fracture of the upper (smaller) chord or tension separation of the rivets over most of the beam and shear separations of the aluminum rivets near the aft end of the forward piece of the beam. The steel bolts between the keel beam and the forward spar were separated when the front spar rotated forward (see section 4.11). The forward rotation of SWB#3 fractured the bathtub fittings before downward motion of the forward end of the keel beam completed fracture of these bolts (see section 4.10). The tension bolts at SWB #2 are failed in pure tension (threads stripped inside nuts). The tension bolts at the midspar failed in tension with the remaining bolts protruding over the keel beam upper chord and bent sharply in the aft direction, consistent with forward motion of the upper edge of the keel beam as the forward end moved downward (pivoting about the last point of fracture, which was the lower chord).

In summary the sequence indicated by the above features is as follows:

5.1.1 SWB#3 rotates forward separating the keel beam tension bolt fittings for this beam and generating about 0.65 inch free play in the joint.

5.1.2 SWB#3 impacts the front spar, causing buckling of the front spar stiffeners, separation of the front spar upper chord in the fillet radius of the chord, and tension separation of the keel beam tension bolts for the spar.

5.1.3 The keel beam is now effectively cantilevered off of SWB#2.

5.1.4 Downward loading on the front of the keel beam from fuselage piece LF6A and associated pieces (see section 7.0 for a more detailed discussion) causes the keel beam to peel away the attachments to the WCS lower skin panel, failing the tension bolts at SWB#3, SWB#2, and the midspar.

5.1.5 As the separation of the keel beam attachments progresses aft, the bending strength of keel beam is exceeded by the continually increasing bending moment causing the keel beam to separate midway between the midspar and SWB#1.

5.2   Overall Wing Center Section (WCS) Sequence

5.2.1   Background of WCS Sequence Development

The overall WCS breakup sequence and early departure of selected parts from the airplane must have been a very precisely orchestrated sequence involving not only the WCS but also the fuselage red area and the keel beam. The sequence integration with the keel beam has been discussed in some detail in Section 5.1 above. More detailed discussions supporting the WCS scenario were provided in Sections 4.1 to 4.11. A more complete integration of the WCS, keel beam, and fuselage red area will be provided in Section 9.0.

5.2.2   Overview of WCS Breakup Sequence

5.2.2.1   There are indications of an early overpressure event (see Section 5.2.3) occurring as far aft as the forward side of SWB#1 and as far forward as the aft side SWB#3 (then front spar after collapse of SWB#3).

5.2.2.2   The spanwise fracture along the upper chord and subsequent forward rotation of SWB#3 due to an overpressure may have been one of the earliest events.

5.2.2.3   SWB#3 impacted the back of the front spar which initiated multiple failures within the spar, setting the stage for lower bulkhead failure, fuselage fracture initiation, and forward keel beam overload.

5.2.2.4   SWB#2 either failed as a result of overpressure, or as a result of the downward separation of the keel beam, or as a combination of these two factors.

5.2.2.5   The WCS maintained wing bending continuity with the upper and lower panels mostly undamaged and the midspar, SWB#1, and rear spar still providing shear continuity. The main landing gear beams also assisted in carrying wing bending.

5.2.2.6   Some localized areas of fire and soot were sustained subsequent to initial events and prior to major airplane breakup (see Section 5.2.4).

5.2.2.7   At major airplane breakup the WCS failed in a manner consistent with up bending overload (the upper panel buckling in compression and the lower panel fracturing in tension).

5.2.2.8   During major airplane breakup the remaining WCS separated with some of the WCS structure remaining attached to the right wing and some remaining attached to the left wing (as described in sections 4.1 to 4.11).

5.2.2.9   WCS structure associated with the right wing became very heavily sooted as a result of a major fire after major airplane breakup.

ZONE _____

COMPONENT _____ UPPER SKIN

SEQUENCE ID NO. ___ US-1 ____

PART ID NO. _____

DESCRIPTION OF POSSIBLE SEQUENCE	SUPPORTING OBSERVATIONS	NON-SUPPORTING OBSERVATIONS	Confidence Level
The loss of the front spar and spanwise beam #3 attachment to the upper skin panel would result in some loss of compression stability of the upper skin panel.			High
If the structure sustains thermal damage from fire, it will result in decreased structural capability.			High
Buckling of the upper skin is potentially initiated between the midspar and the rear spar and restraint would be provided by the BL 0 rib which is consistent with an inflection in the curvature at BL 0.	Upper skin panel has residual deflection of an upward bow to the left of BL 0 and slight upward curvature to the right of BL 0 with the inflection point at BL 0.		Medium
The initial fracture consistent with upper panel buckling is potentially along the attachment to spanwise beam #1. This fracture runs primarily through the single fastener row common to the upper chord of the spanwise beam.	1. Bending fracture in the skin panel at SWB #1 is through the fastener row and the fracture exhibits upward bending both fwd and aft of the fastener row. 2. Fore/aft fractures are not continuous forward or aft of the fracture running along SWB#1 suggesting that the fracture along SWB#1 existed prior to the fractures occurring at LBL 40, LBL 5, RBL 76, and RBL 100. (see figure B-1)		Medium

01/24/97

204

APPENDIX IV

Overview of Cooperative Engagement Capability (CEC)

The secretary of defense has said that Cooperative Engagement Capability (CEC) is the biggest breakthrough in warfare technology since Stealth. The CEC program is designed to link together Battle Group Anti-Air Warfare Units and Airborne Early Warning aircraft into a force-wide antiair combat system. CEC provides real-time, high-quality, composite data over highly jam-resistant links.

> John W. Douglas
> Assistant Secretary of the Navy
> March 29, 1996

The CEC program actually came into existence in 1988 "under the auspices of the Battle Group Anti-

Air Warfare Coordination program." In 1990, tests were conducted at sea. By 1992, the program had reached a sufficient level of maturity to rapidly accelerate. Acquisition of software and hardware began in 1992 and by early 1994, "after a series of preliminary trials, five CEC-equipped ships, including the amphibious assault ship *Wasp* and the aircraft carrier *Dwight D. Eisenhower,* as well as P-3 Orion maritime aircraft, verified the ability of CEC units"[1] to independently "construct identical composite tracks and identification pictures"[2] by remotely linking their radar software. In other words, the Navy was fairly sure the theory would work.

This new system would be tied together via Cooperative Engagement Capability, laboriously described in 1994 as a system that will significantly enhance capabilities in joint theater air and self-defense missions against reduced signature cruise and theater ballistic missiles by combining tracks from dispersed force censors into a real-time, accurate, fire-control-quality Anti-Air Warfare (AAW) picture shared force-wide. Cooperative Engagement's high data rate and real-time exchange of fire control sensor data will greatly expand mission effectiveness in the littoral."[3] Littoral literally means between the low and high water marks.

[1] David Foxwell, Jane's International Defense Review, Technical Feature, Tasks and Threats Multiply for Amphibious Forces, no date, p. 2.
[2] Ibid.
[3] Ibid.

"Today, since no nation can challenge our ability to control the seas, we have concentrated our planning on winning the contest for control of the land and sea areas of the littoral," the 1994 Navy publication began.[4] Generally, "littoral" warfare covers the area from the shore to the open sea, and "inland from the shore over that area that can be supported and controlled directly from the sea."

The 1994 littoral concept envisioned fully integrated joint operations with the Army and Air Force as well as with allied forces. Now that the Cold War is over, there is a shifting concept to a much greater emphasis on fighting on land rather than over vast stretches of the ocean. Isolated Naval missions take a backseat to the need for a "seamless" integration of the fighting equipment, particularly the software that operates the sophisticated AEGIS radar system.

Desert Storm, in 1991, clearly demonstrated that "the proliferation of theater ballistic missiles (TBMs) poses increasing danger to the national security of the United States and our allies."[5] So the Navy decided to focus its advanced concept thinking around the AEGIS system. In 1994, the Navy predicted that "In the near future, AEGIS cruisers and Arleigh Burke (DDG 51) destroyers will provide a somewhat limited, but nonetheless highly mobile and credible theater

[4] Navy Public Affairs Library, Department of the Navy 1994 Posture Statement.
[5] Ibid.

209

ballistic missile defense (TBMD) capability. When AEGIS SPY-1 radar software improvements are combined with improvements to the Standard missile, these ships can provide lower tier defense against incoming ballistic missiles."[6]

By 1995, the Navy's CEC concept was more clearly defined, not to mention more readable: "CEC is a system of hardware and software that allows the sharing among ships of radar data on air targets. Radar data from individual ships of a battle group is transmitted to the other ships in the group via a line-of-sight data distribution system. Each ship uses similar data-processing algorithms resident in its cooperative engagement processor, resulting in each ship having essentially the same displays of track information on aircraft and missiles. An individual ship can launch an antiair missile at a threat aircraft or an antiship missile within its engagement envelope, based on track data relayed to it by another ship."[7]

"To augment these capabilities and provide over-the-horizon early warning, we have embarked on a joint program with the Army to develop and field Joint Tactical Ground Stations (JTAGS). JTAGS will allow in-theater processing of space-based warning data, greatly enhancing the abilities of active theater defense." And all the high-tech equipment will work with that being developed by the other services. This

[6]Ibid.
[7]DoD, FY 95 Annual Report, Cooperative Engagement Capability.

level of cooperation had been talked about over the decades, but this was the first time it appeared that true cooperation would exist. Shrinking budgets and the demands of 21st-century warfare finally overpowered interservice rivalry.

Then the Navy correctly identified the problem that would lead to the Flight 800 tragedy two years later:

Congestion in littoral war zones combined with the complexities of the sea, air, land, and space interfaces will increase the difficulty of identifying and sorting the dispositions of friendly, neutral, and hostile forces. Doing so has become increasingly critical as weapon lethality has increased and target engagement response times have decreased. Enhancements to the current Position Location Reporting System and increased fielding of the Global Positioning System have provided greater capability for the positive identification of friendly ground forces.[8]

The P-3 aircraft that would play a role in the July 17, 1996, tragedy were also due for upgrade in order to play a role in the littoral warfare future: "In particular, we are improving the surveillance systems of the P-3 to make it more useful in the missions we now envision. Upgrades include addition of long-range optical systems, radar upgrades, and improved command and control systems."[9]

[8] Ibid.
[9] Ibid.

In September 1995 tests were conducted in the Gulf of Mexico. Two CEC-equipped AEGIS cruisers and an "airborne early warning (AEW) P-3 aircraft owned by the Coast Guard," were used to test the "effectiveness and suitability of an airborne CEC."[10]

The Department of Defense noted that its 1995 accomplishments included completing "analysis of Developmental Testing/Operational Testing (DT/OT) lessons learned to fully support continued developmental efforts in CEC system design and fleet operations and tactics. . . . [And they] Continued development of airborne CES for integration with E-2C aircraft."[11] Also accomplished was the "modeling and simulation of ship-based over-the-horizon cruise missile defense with airborne surveillance and tracking to develop operational concepts for deployment jointly with the Army and Air Force. [And] Work to design a system to transfer Cooperative Engagement Capability (CEC) data to [an] Army Patriot battery for analysis of future development and in preparation for simulated Army missile-firing events."[12]

Just as the early 1996 CEC tests were getting underway off the island of Kauai, Undersecretary of Defense for acquisition and technology, Paul G. Kaminski,

[10] Ibid.
[11] DoD FY 97 Descriptive Summary.
[12] Ibid.

gave a speech at the Redstone Arsenal in Huntsville, Alabama:

> "We have seen equally encouraging field demonstrations of the Navy's Cooperative Engagement Capability, which has been deployed in TMD [theater missile defense] exercises with the [aircraft carrier] *Eisenhower* battle group off the Atlantic coast and in the Mediterranean over the course of the past twelve months as part of the JTF-95 [joint task force] exercise activity."

So CEC testing was not limited to the sterile environment of the missile range off Kauai.

During phase one of development, the focus of the cruise missile defense was "the detection and engagement of beyond-the-radar-horizon cruise missile targets. The goal was to detect, track, and successfully engage cruise missiles at ranges beyond the radar line of sight of surface-based air defense units."[13] Sensors were placed on a Hawaiian mountaintop, giving the altitude to simulate an aircraft acting in concert with an AEGIS cruiser and a U.S. Army Patriot battery "to detect, track, and engage target drones at ranges beyond the radar lines of sight of the surface-based air defense units."

This was a joint Navy/Army testing and develop-

---

[13]CINCPAC, ACTD Master plan Cruise Missile Defense Phase I.

ment program, with Navy lead. The scenario resembled what the military planners saw as the limited war capability in the remote hot spots of the world. Two AEGIS cruisers from the Atlantic fleet participated, the *Anzio* and the *Cape St. George,* as well as the carrier *Eisenhower,* a Customs Service CEC-equipped P-3, an Army Patriot missile battery and a USAF E-3A AWACS aircraft, a Marine Corps Hawk missile battery, and a sensor-equipped aerostat.

Several air defense scenarios were run in January and February of 1996 off the Hawaiian islands at the Kauai Pacific Missile Range Facility, in order to test the rapidly developing high-tech program. As confidence built, the scenarios began to be tested "in jamming and radar-clutter environments."[14]

Instead of using expensive cruise missiles, the Navy's target of choice while developing CEC has been the BQM-74E drone produced by Northrop Grumman. Configured to closely resemble a cruise missile, the BQM-74E can be remotely controlled or preprogrammed to fly a specific route. An aircraft resembling a BQM-74E drone would be photographed later in the same area, during the same time frame, where Flight 800 was shot down. Where the BQM-74E drone flies, things are being shot into the air.

Nine of these drones were used during the Army/

[14]CINCPAC, Cruise Missile Defense Phase 1.

214

Navy test of this CEC concept. All were destroyed. Two AEGIS cruisers used CEC to defend themselves, firing Standard missiles to destroy the drones. "The ships also communicated with a nearby Army Patriot antimissile radar and AWACS"[15] in order to successfully engage and destroy the BQM-74E drones.

An Army unit was on the eastern end of Long Island in the days leading up to the shootdown of TWA Flight 800, engaged in military exercises that included the launch of several drones.[16]

The USS *Lake Erie*, one of the AEGIS cruisers involved in the January 20 and 21, 1995, tests off Kauai, used a Standard missile, "modified for remote engagements, to kill the BQM-74E drones, which were flying out of radar range at an altitude of fifty feet."[17]

The second set of tests, February 1 and 2, 1996, "were far more complex . . . These exercises added Army and Air Force assets"[18] as well as a CEC Customs Service P-3. Two additional AEGIS cruisers, the USS *Anzio* and *Cape St. George,* were joined with the USS *Lake Erie.*

The scenario used for the test included the AEGIS vessels' arrival in a hostile littoral environment. "The

[15]John Donnelly, Defense Week, Joint Exercises Establish Firsts For Cooperative Engagement, February 20, 1996, p. 6.
[16]Sensitive source.
[17]Ibid.
[18]John Donnelly, Defense Week, February 20, 1996, p. 6.

battle group was confronted with two drones, launched from the island."[19] Each drone employed its own "self-screening" jammers as they streaked toward the AEGIS battle group, less than fifty feet above the ocean. The BQM-74E drone jammers "blinded all the cruisers. The *Lake Erie* alone regained the track in time."[20] Because the radar data is shared by all CEC-equipped AEGIS ships, even the *Anzio* and *Cape St. George* computer systems could track the oncoming drones. Each was able to launch a Standard missile and destroy a drone.

All the AEGIS radars being simultaneously blinded should have been sufficient warning that this system could not be tested in any environment close to civilian air traffic. If the Standard missile had been launched just before all the AEGIS radars were jammed, the Standard missile, with its semiactive radar, would have been on its own to continue to climb into the sky, its radar searching for a target.

One month later, the Secretary of the Navy, John H. Dalton, testified before the Senate Armed Services Commitee:

A second successful investment in emerging technologies is our Cooperative Engagement Capability, or CEC. Beginning with highly successful live firing tests in the summer of 1994

[19] Ibid.
[20] Ibid.

and continuing through a series of challenging demonstrations and exercises in the past year, CEC continues to exceed our most optimistic expectations.

Most recently, CEC was a key element in the Advanced Concept Technology Demonstration, better known as Mountain Top, which took place in Hawaii last month. In Mountain Top, the Navy proved that it can conduct surface-to-air engagements of cruise missiles while those threats are still located far beyond the ships' own radar location.

The true significance of Mountain Top is that our surface combatants will have the capability to provide effective air defense of forces ashore, debarkation ports, and airfields against low-flying, Tomahawk-like cruise missiles. Secretary Perry has declared CEC the most significant technological development since Stealth."[21]

Secretary Dalton's emphasis on the Navy's ability to protect Army and Marine forces onshore should not be minimized. Future testing of CEC as it continued its march toward combat certification, required an increasing level of testing providing a realistic setting that included Army and Navy CEC assets working in conjunction, drone missiles launched in land clut-

[21] Secretary of the Navy, John H. Dalton, statement before the Senate Armed Services Committee, March 12, 1996.

ter, and an ever-increasing level of jamming, until the system could be proven to successfully function in a realistic combat environment, all the while maintaining the ability to distinguish military from civilian aircraft while simultaneously locating a drone coming out of land clutter in the same area as the civilian and military planes.

And, as a Navy document explained shortly before Flight 800 was shot down: "The Navy also has begun a study to investigate the difficulties inherent in shipboard sensors in littoral environments." Shipboard sensors are synonymous with AEGIS radar. Littoral means land in the vicinity of which a cruise missile attack can be launched. All these factors were present in the days leading up to the shootdown of Flight 800. An Army unit is reported to have been involved in a series of BQM-74E drone launches from a position on eastern Long Island, a necessary condition for a Navy study of the "difficulties inherent in shipboard sensors in littoral environments."

The same document warns of Antiship Cruise Missiles (ASCMs): "Increasingly available throughout the world, these sophisticated, relatively inexpensive weapons can be launched from the air, sea, or land. The limited time available to react to them, once airborne, could pose difficulties for existing antiair defenses, particularly in littoral operations where naval forces may be patrolling very close to the shore or in physically constrained bodies of water. A num-

218

ber of countries in regions vital to American interests, including the Gulf, now possess advanced ASCMs."[22]

The Department of Defense 1996 plan was to complete "Initial Operational Capability (IOC) certification for the shipboard system. . . . Continue development of airborne CEC for integration with E-2C aircraft. . . . [and] Modify Naval Research Laboratory (NRL) and fleet-owned P-3 aircraft to provide dedicated airborne support for CEC test programs."[23] In other words, when CEC was being tested, there would be a modified P-3 in the air monitoring the test. There was a modified P-3 in the air almost directly above Flight 800, monitoring a test when the 747 was shot down.

By 1996, CEC was headed for its initial certification for use by the Navy in combat. The CEC development project was now described as, "coordinating all Battle Force sensors into a single, real-time, composite track picture having fire-control quality."[24]

While tedious, the military definition of CEC in its advanced stage of development is essential to understanding the complex system we believe failed at a critical moment during one of its final tests prior to combat certification: "CEC distributes censor data from each ship and aircraft, or cooperating unit (CU), to all other CUs in the battle force through

[22]Maritime Forces, Chapter 19.
[23]DoD FY 97 Descriptive Summary.
[24]DoD FY 97 Descriptive Summary.

219

a real-time, line-of-sight, high-data-rate sensor and engagement data distribution network. CEC is highly resistant to jamming and provides accurate gridlocking between CUs. Each CU independently employs high-capacity parallel processing and advanced algorithms to combine all distributed sensor data into a fire-control-quality track picture which is the same for all CUs. CEC data is presented as a superset of the best AAW sensor capabilities from each CU, all of which are integrated into a single input to each CU's combat weapons system. CEC will significantly improve our Battle Force in depth, including both local area and ship defense capabilities against current and future AAW threats. CEC is designed to enhance the AAW war fighting ability of ships and aircraft and to enable coupling of the Force into a single, distributed AAW weapon system and toward more effective use of tactical data and the cooperative use of all the Force sensors and weapons. These capabilities will provide the ship defense flexibility needed to meet the threat brought about by increasing numbers of highly sophisticated weapons held by potentially hostile third-world countries.

"CEC consists of the Data Distribution System (DDS), the Cooperative Engagement Processor (CEP), and Combat System Modifications. The DDS encodes and distributes onship sensors and engagement data, is a high-capacity, jam-resistant, directive system providing a precision gridlocking and high throughput of data. The CEP is a high-capacity distributed processor which is able to process force levels of data in a timely manner that allows its output to be considered real-time fire-control data. This data is passed to the ship's combat system as fire-control quality data for which the ship can cue its onboard sensors or use data to engage targets without actually tracking them."[25]

A Pentagon document, in typical military techno-jargon, described the test CEC would have to success-fully pass before being certified for combat: "The Cruise Missile Defense (CMD) Advanced Technology effort includes: an Advanced Concept Technology Demonstration (ACTD), Phase 1, which demon-strates that an AEGIS ship (or other surface-based missile launch platform), using one or more surro-gate airborne sensor partners, can provide greatly expanded air defense capabilities leading to a robust capability against overland cruise missiles beyond surface-based radar line-of-sight."[26]

The Naval Surface Warfare Command (NSWC), East Coast Operations (ECO), located at Dam Neck, Virginia, "was selected as the CEC test site due to its existing hardware resources and physical location with respect to the Virginia Capes surface ship operating area. Particular existing NSWC ... hard-

---

[25] Ibid.
[26] DoD FY 1997 Descriptive Summary.

ware assets being used for the CEC test site include . . . SPS-48E air search radar, SPS-49V5 search radar, SPS-48C search radar . . . air intercept control facility, and test control central . . . ECO (East Coast Operations) NSWC (Naval Surface Warfare Command) is located on the shore directly adjacent to a live gunfire range and the Virginia Capes Operating Area. This arrangement facilitates tests of tactical computer programs with live shore-based radars and communication equipment with ships and aircraft operating at sea. The actual operational environment provides extremely valuable test data that is used to upgrade the tactical computer programs to meet fleet needs. Additionally, ECO NSWC employs a direct microwave link to the FACSFAC air search radars to provide live radar displays for training and testing. ECO radars act as backups in the event of problems with the FACSFAC radars."[27]

ECO NSWC was also responsible for:

—CEC land-based test site management.

—Air logistic support of CEC battle group in Virginia Capes operating area.

So the Naval Surface Warfare Command, East Coast Operations, located at Dam Neck, Virginia, had the technical expertise to closely monitor all CEC activity in the restricted zones and warning zones that extend from Virginia to just south of Long Island. These tests are monitored and tape-recorded for training

[27] ECO SSA Technical Narrative.

purposes. And aircraft are in the air with monitoring equipment to enhance the collection of data.

This monitoring equipment recorded the events that led to the destruction of Flight 800, a few miles northwest of active warning zone W-105. The only question is: have the tapes been destroyed?

We know that CEC was in the final testing phase prior to being certified for combat. The test bed for the final months was the area between Long Island and Virginia. The monitoring system for the tests is headquartered at Dam Neck, Virginia. Aircraft outfitted with monitoring equipment were under their control. The entire radar monitoring system was even plugged in to the military version of the FAA, at Oceana Virginia, FACS/FAC. These military air controllers monitor military traffic in the area where CEC is tested.

After Flight 800 went down, nothing more was heard of the need to test CEC's ability to operate with an air asset, in land clutter as well as when being jammed. There was no longer a need to have friendlies and commercial aviation on the screen being read by the AEGIS computer. A kinder, gentler CEC testing program surfaced far away from civilian traffic and a mere shadow of itself in the recent glory days, when overconfidence prevailed.

APPENDIX V

DECEMBER 26, 1996
LETTER FROM
BERNARD S. LOEB,
DIRECTOR OF
AVIATION SAFETY,
NTSB, TO DAVID F.
THOMAS, DIRECTOR
OF ACCIDENT
INVESTIGATION, FAA

National Transportation Safety Board

Washington, D.C. 20594

Mr. David F. Thomas
Director
Office of Accident Investigation (AAI-1)
Federal Aviation Administration
Washington, D.C. 20591

December 26, 1996

Dear Mr. Thomas:

The purpose of this letter is to request a written explanation of certain events related to the processing of Federal Aviation Administration air traffic control (ATC) radar data for TWA flight 800 that crashed near Long Island, New York on July 7, 1996. As we have discussed, I would like to clarify the circumstances to alleviate any potential future misunderstandings or inappropriate speculation regarding the results of preliminary radar data analyses.

As you know, during the first few hours after the accident, some FAA personnel made a preliminary assessment that recorded ATC radar data showed primary radar hits that indicated the track of a high speed target that approached and merged with TWA 800. One of your staff called our office about 0930 on July 8, 1996, to advise us of the preliminary assessment of the radar data by FAA personnel, suggesting that a missile may have hit TWA 800. This preliminary assessment was also passed to other government officials, including White House officials. After the Safety Board received the ATC radar data and reviewed it, it was determined that the preliminary assessment by FAA staff was incorrect. We understand that FAA officials now agree with the Safety Board's determination.

I would appreciate it if you could verify that all specialists and/or managers involved in the preliminary radar analyses fully agree that there is no evidence within the FAA ATC radar data of a track that would suggest a high speed target merged with TWA 800. I would also appreciate an explanation about how the preliminary incorrect assessment occurred, so that potential public or media inquiries can be handled in a accurate and consistent manner.

If you have any questions about this matter, please call me or Mr. Ron Schleede. I trust that you appreciate the need to ensure a clear record of these particular events to allay public concerns or speculation.

Bernard S. Loeb

APPENDIX VI

LETTER IN REPLY FROM DAVID F. THOMAS, FAA, TO BERNARD S. LOEB, NTSB

U.S. Department
of Transportation

Federal Aviation
Administration

800 Independence Ave., S.W.
Washington, D.C. 20591

Dr. Bernard S. Loeb
Director of Aviation Safety
National Transportation Safety Board
490 L'Enfant Plaza, SW.
Washington, DC. 20594-2000

Dear Dr. Loeb: ~Bernie~

This is in response to your letter of December 26, 1996,
regarding the Federal Aviation Administration's (FAA)
processing of preliminary radar data following the TWA
Flight 800 accident on July 17, 1996.

During the night of the accident, one of the many concerns
of FAA air traffic personnel was the possibility of a midair
collision between two aircraft. In an attempt to conduct a
rapid assessment of this possibility, personnel at the
New York Air Route Traffic Control Center (ZNY) replayed
the ZNY radar data at the facility using a commercially
available radar replay software program called "Radar
Viewpoint." The review of the printout from the program
indicated that there were radar tracks which could not be
accounted for by FAA staff. This information was
immediately relayed to the appropriate law enforcement
organizations with the understanding that it was preliminary
and did contain some unexplained data.

Subsequently, after receiving the request for radar data
from the National Transportation Safety Board (NTSB), all
radar information from every radar site which had recorded
information on TWA 800 was provided to the NTSB.
Concurrently, an exhaustive internal review of those data
was conducted at the FAA Technical Center. The assessment
by the FAA Technical Center indicated that the likelihood
of a missile was remote. It must be noted, however, that
FAA air traffic radar is designed to detect and monitor
aircraft, not high-speed missiles, so any conclusions based

on this review must consider the technical limits of the radar. Since that time, there has been no other evidence developed from the radar data that would indicate the existence of a missile.

Your letter asks the FAA to "verify that all specialists and/or managers involved in the preliminary radar analyses fully agree that there is no evidence within the FAA ATC radar data of a track that would suggest a high speed target merged with TWA 800." Although we understand and share your desire to allay public concern over this issue, we cannot comply with your request. The ZNY facility personnel in question do not possess the indepth technical background required to conduct the level of analysis needed to positively reach a conclusion on the significance of the radar data. The preliminary assessment made by ZNY facility personnel on the night of the accident was as thorough as possible but was, and is, limited by technical factors. Therefore, they would neither agree nor disagree with that assessment.

Regarding the notification of the White House and other Government officials, you will recall that immediately after the event there was speculation within the media and other organizations of possible terrorist activity. By alerting law enforcement agencies, air traffic control personnel simply did what was prudent at the time and reported what appeared to them to be a suspicious event. To have done less would have been irresponsible.

I trust that this information is responsive to your concerns. If I can be of further service, please let me know.

Sincerely,

David F. Thomas
Director of Accident Investigation

APPENDIX VII

U.S. DISTRICT COURT GRAND JURY SUBPOENA, DELIVERED BY FBI 3/24/97

F.#9603237

United States District Court

Eastern _____ DISTRICT OF _____ New York

TO: Zebra Books
c/o Paul Zinas
850 Third Avenue - 16th Floor
New York, New York 10022

**SUBPOENA TO TESTIFY
BEFORE GRAND JURY**

SUBPOENA FOR:
☐ PERSON ☒ DOCUMENTS OR OBJECT(S)

YOU ARE HEREBY COMMANDED to appear and testify before the Grand Jury of the United States District Court at the place, date, and time specified below.

PLACE	ROOM
United States District Court Eastern District of New York 225 Cadman Plaza East Brooklyn, New York 11201	478
	DATE AND TIME
	April 7, 1997

YOU ARE ALSO COMMANDED to bring with you the following document(s) or object(s):*

Please provide any and all documents relating to any book or publishing contract for James Sanders for "The Downing of TWA Flight 800" including but not limited to any contracts, draft contracts, correspondance, offer letters, payment records, cancelled checks or check stubs, telephone or e-mail messages, and documents reflecting the date negotiations began and when the contract was signed, finalized or concluded.

YOU MAY COMPLY WITH THIS SUBPOENA BY PROVIDING THE REQUESTED DOCUMENTS TO SPECIAL AGENTS JIM KINSLEY OR ANTHONY JACKSON OF THE FEDERAL BUREAU OF INVESTIGATION AT (516) 753-0130.

☐ Please see additional information on reverse

This subpoena shall remain in effect until you are granted leave to depart by the court or by an officer acting on behalf of the court.

CLERK	DATE
Robert C. Heineman	3/24/97
(BY) DEPUTY CLERK	

This subpoena is issued upon application of the United States of America	NAME, ADDRESS AND PHONE NUMBER OF ASSISTANT U.S. ATTORNEY Benton Campbell, AUSA One Pierrepont Plaza - 19th Floor Brooklyn, New York 11201 (718) 254-6384

*If not applicable, enter "none."   To be used in lieu of AO110   FORM OBD-227 JAN 86

BIBLIOGRAPHY

Documents

Department of the Army, Force XXI Campaign Plan.

Department of the Army, Tradoc Pamphlet 525-5. Force XXI Operations.

CINCPAC, ACTD Master Plan, Cruise Missile Defense, Phase 1.

CINCPAC, Cruise Missile Defense, Phase 1.

Secretary of the Navy, John Dalton, statement before the Senate Armed Services Committee, March 12, 1996.

DOD, FY 95 Annual Report, Cooperative Engagement Capability.

DOD, FY 97, Descriptive Summary.

ECO SSA Technical Narrative.

Executive Order 12958.

Ken Fuhrman, Application of VME Hardware on the Cooperative Engagement Capability (CEC) Program, no date.

National Transportation Safety Board, NTSB, Sequencing Team Final Report.

NTSB/Coroner color-coded study matching body damage to seat damage.

NTSB, TAGS Database, Debris field, Red, Yellow, Green Zone, November 1996.

NTSB, *TWA 800, Chairman's Briefing/Status Report*, November 15, 1996.

NTSB, Metallurgy/Structures Sequencing Group Report, Appendix B: Detailed Rationale.

U.S. Navy, Maritime Forces, Chapter 19.

Navy Public Affairs Library, Department of the Navy 1994 Posture Statement.

West Coast Analytical Service, Inc., three-page analysis of residue report, Chain of Custody Record.

See documents in the appendix.

Newspapers/Magazines

Al Baker, *Newsday*, "Missile Theory is Still Aloft," September 1, 1996.

Al Baker, *Newsday*, "Kallstrom Keeps Up the Quest for TWA Clues," September 24, 1996.

Susan Candiotti, *CNN*, Bob Francis interview, August 22, 1996.

Bibliography

Rita Ciolli, *Newsday*, "Information Instead of News," July 23, 1996.

CNN, "NTSB Confirms Fuel Tank Exploded," August 25, 1996.

John Donelly, *Defense Week*, "Joint Exercises Established Firsts for Cooperative Engagement," February 20, 1996.

William B. Falk, *Newsday*, "Fed: Help Us," July 21, 1996.

Loren Fleckenstein, *The Riverside Press-Enterprise*, "Debris Patterns Provide Key," March 10, 1997.

David Foxwell, *Jane's International Defense Review*, Technical Feature, Tasks and Threats Multiply for Amphibious Forces, no date.

Craig Gordon and Knute Royce, *Newsweek*, "Agencies Vie for Control After Flight 800's Demise," August 11, 1996.

David E. Hendrix, *The Riverside Press-Enterprise*. "New Data Show Missile May Have Nailed TWA 800," March 10, 1997.

David E.Hendrix, *The Riverside Press-Enterprise*. "Solid Fuel for Rockets Follows Basic Recipe," March 10, 1997.

David E. Hendrix, *The Riverside Press-Enterprise*, "FBI Impounds Radar Image Tape," March 11, 1997.

David E. Hendrix, *The Riverside Press-Enterprise*, "Missile Fragment Theory," March 12, 1997.

David E. Hendrix, *The Riverside Press-Enterprise*, "FBI, Pentagon Criticize Reports on TWA," March 12, 1997.

IFFA, Flight Attendant Advocate, Special Memorial Edition 1996, Volume 2, Issue 2.

Reed Irvine, *Accuracy in Media*, March 17, 1997.

Robert E. Kessler, *Newsday*, "Missile Tests in Flight 800 Probe," December 11, 1996.

Earl Lane, *Newsday*, "Support for Tank Explosion Theory," September 21, 1996.

Earl Lane & Knute Royce, *Newsday*, "Calling in More Experts," August 17, 1996.

Pat Milton, *Associated Press*, "Video May Show Missile Near TWA . . . ," March 11, 1997.

Pat Milton, *Associated Press*, "FBI Acknowledges Missile Theory," March 13, 1997.

Pat Milton, *Associated Press*, "TWA Crash Theorist May be Charged," March 13, 1997.

Pat Milton, *Associated Press*, "FBI Explains TWA 800 Radar Blip," March 21, 1997. Follow up, March 26, 1997.

Pat Milton, *Associated Press*, "TWA Friendly Fire Theory Out," September 16, 1996.

Newsday, "Searching for Answers," July 19, 1996.